SPACE AGE ADVENTURES

SPACE AGE ADVENTURES

OVER 100 TERRESTRIAL SITES AND

OUT OF THIS WORLD STORIES

MIKE BEZEMEK

UNIVERSITY OF NEBRASKA PRESS | LINCOLN

The University of Nebraska Press is part of a land-grant institution
with campuses and programs on the past, present, and future
homelands of the Pawnee, Ponca, Otoe-Missouria, Omaha,
Dakota, Lakota, Kaw, Cheyenne, and Arapaho Peoples, as well as
those of the relocated Ho-Chunk, Sac and Fox, and Iowa Peoples.

Library of Congress Cataloging-in-Publication Data
Names: Bezemek, Mike, author.
Title: Space age adventures: over 100 terrestrial sites
and out of this world stories / Mike Bezemek.
Description: Lincoln: University of Nebraska Press,
2023. | Includes bibliographical references.
Identifiers: LCCN 2022043566
ISBN 9781496230829 (paperback)
ISBN 9781496236524 (epub)
ISBN 9781496236531 (pdf)
Subjects: LCSH: Astronautical museums—United States—
Guidebooks. | Aerospace industries—United States—Guidebooks. |
Historic sites—United States—Guidebooks. | Astronautics—
United States—History—Sources. | BISAC: HISTORY / United
States / 20th Century | TRAVEL / United States / General
Classification: LCC TL506.U5 B49 2023 | DDC
629.130074/73—dc23/eng/20221109
LC record available at https://lccn.loc.gov/2022043566

Set and designed in Lyon Text and Neutraface 2 Text by N. Putens.

Frontispiece-a. An artistic depiction of exoplanet Kepler-16b, which was discovered by the Kepler Space Telescope. NASA/JPL.

Frontispiece-b. Saturn IB and Saturn V rise above the U.S. Space & Rocket Center in Huntsville AL. Mike Bezemek.

CONTENTS

PREFACE
Space Sites and Stories

When people dream about space travel, they naturally tend to look up. After all, some of the greatest adventures in human history happened high above the planet, in outer space. But there are many places down here on Earth where enthusiastic observers can have their own space age adventures. In fact, since the United States has been a central player during the first century of space exploration, over a hundred relevant places can be conveniently found across the country.

Before Apollo astronauts walked on the Moon, they trained at sites that you can still visit today—from NASA space centers and telescope observatories to impact craters and atomic testing grounds. Inside vast museum hangars, you can walk beneath a towering Saturn V rocket left over from the Apollo program, or peer inside a space-flown capsule, like those from Projects Mercury and Gemini, with heat shields singed by reentry. Elsewhere, you might visit historic rocket pads or even watch a scheduled launch.

This book presents over one hundred terrestrial adventures related to the space age that you can experience right here on Earth. Most of these trips are organized into regional clusters within the lower forty-eight states. Among these trips, special attention is paid to space-flown artifacts, life-sized replicas, immersive experiences, and exciting locations like astronaut training sites. The goal is to help enthusiasts connect with the exhilarating endeavors that continue to unfold among the stars.

Of course, objects alone can feel somewhat lifeless when you don't know the adventurous stories behind the artifacts. For that reason, in addition to trips, this book presents a multipart series of true stories that span the

1. Earth is captured by an automatic camera during the uncrewed *Apollo 4* mission. NASA.

space age. Within these stories, special attention is paid to both historically significant episodes and lesser-known or forgotten exploits. Plus, effort has been made within the stories to place many important space artifacts, like those you might find in a museum, into the larger context of what happened during the mission. The result is a condensed story of the space age designed to help readers develop a deeper understanding of the history and drama behind the historic artifacts and famous images.

With this book as your guide, you'll not only see the top sites but discover the unforgettable tales as well. Along the way, you will bring to life one of the most dramatic eras in human exploration.

ACKNOWLEDGMENTS

A big thank you to everyone who helped in the creation of this book. To my wife, Ina Seethaler, for joining me on many trips to space sites across the country. To the many insightful museum people I've met along the way. To early reviewers Jay Gallentine and Chris Gainor. And to the excellent team at the University of Nebraska Press who made this book possible, including Rob Taylor, Courtney Ochsner, Rosemary Sekora, Ann Baker, Nathan Putens, Stephanie Marshall Ward, and the rest of the staff. Thanks so much!

Dedication: For the best friends on Earth, including Guido Greg PE, B. Mo. Braaaaaaad, Levon the Luh-vawn, Lee pahk-the-cahr J, and Boberts Hawaii MD. I look forward to our space jalopy.

2. The Transonic Wind Tunnel at NASA's Langley Research Center. NASA.

BECOMING A SPACE AGE ADVENTURER

It felt like driving across another planet. Volcanic cones and rolling hills, wide plains pocked by basins and swales. In all directions, this mysterious surface was covered by crumbly black cinders that crunched beneath my tires. The overlapping 4x4 tracks and occasional stands of pine trees were the most familiar terrestrial features.

I was in the Cinder Hills OHV Area in Northern Arizona. I'd spent the past seventeen days rowing the Colorado River through the Grand Canyon. It was my sixth visit to this world wonder and my third multiweek river trip through what American explorer John Wesley Powell called the Great Unknown. But right now, the bigger unknown was a disoriented *where on Earth am I?*

I'd spent the previous night floating under the stars beyond the Grand Wash Cliffs toward Pearce Ferry on eastern Lake Mead. Now I was too tired to make the drive back to my summer base near Moab, so I was searching for a last-minute campsite in the mountains outside Flagstaff.

Beyond my window, one part of the area looked like it had been hit by not just a volcanic eruption but possibly a meteor strike. I didn't learn this fact until I got home, but it turns out the Cinder Hills *were* bombarded in the 1960s. Not by extraterrestrial impactors but by some adventurous specialists, affiliated with the USGS and NASA. They were trying to recreate a portion of the Sea of Tranquility, where *Apollo 11* would land in 1969.

When I was a kid, I was fascinated by space and interstellar exploration. I read magazines like *Air & Space* and science fiction books like Douglas Adams's Hitchhiker series and Arthur C. Clarke's monolith trilogy. Favorite films included *2001*, *Apollo 13*, and the miniseries *From the Earth to the Moon*. My preferences were often directed toward space adventure, like *Star Wars* and *Babylon 5*, especially when mixed with scientific exploration, like *Star Trek*.

After I became a wilderness and whitewater rafting guide in college, my interests shifted to having my own adventures here on Earth. Yes, I still visited space museums and peered inside spacecraft. But when a company began soliciting applications for one-way tickets to Mars, I decided not to apply. By then, the *Mars Reconnaissance Orbiter* had spent over a decade searching unsuccessfully for regularly flowing water on the surface. No Martian rivers to paddle? Abort, thanks.

But reading about the NASA history of the Cinder Hills stoked my interest in rediscovering the space age. I thought back on some of the places I'd already visited. I'd been to a ton of museums, including the Smithsonian Air and Space Museum on a family trip as a kid. During a volcanology summer field camp, I'd toured the mountaintop observatories on the Big Island of Hawaii. I'd viewed from afar the offices of Wernher von Braun at Stennis Space Center, where they test-fired the Saturn V rocket engines. I'd hiked around Upheaval Dome in Canyonlands, which has renewed support among geologists as an impact feature. Even the Grand Canyon had a connection to space exploration, with Neil Armstrong and Buzz Aldrin having hiked the trails to and from Phantom Ranch, studying geology along the way.

This one visit to the Cinder Hills got me curious about all the other earthly places with space age significance that could be visited across the United States. Soon I began ramping up my visits, and not just the typical air and space facilities but remote training sites and obscure museums that somehow got a Moon rock. I walked around retired space shuttles and stared up at towering rockets standing like they did on launch day. Along the way, I discovered some pretty cool trips, with stops ranging from historic rocket ranges to mobile quarantine units to stellar observatories and working spaceports.

As I delved into the project, I became increasingly aware of how many dramatic episodes from the space age have been forgotten by all but the most diehard enthusiasts. The more I researched the stories behind the artifacts and replicas, the more the museum exhibits came to life during visits. This feedback loop deepened my appreciation for not only space missions of the past but space missions of the future as well.

We've arrived at the dawn of a second space race. Public-private partnerships will carry new women and men back to the Moon, onward to Mars, and hopefully beyond. Many of us on Earth will be watching their adventures from afar, which bumps my EKG every time. If you feel the same, having your own space age adventures here on Earth can be the perfect way to connect more deeply with the events soon to unfold above our planet.

This is the part where some might say, *blast off*. Instead of that astronaut knee slapper, may I suggest *T-minus 293 pages*?

3. A SpaceX Cargo Dragon spacecraft approaches the International Space Station in 2021. NASA.

USING THIS BOOK AND MUST-SEE SITES

How to Use This Book

✓ Use this book to plan amazing trips and adventures to the nation's top space museums, science centers, planetariums, observatories, rocket parks, astronaut training sites, and more!

✓ Use this book to discover out-of-this-world true stories from the space age, including historic missions, forgotten episodes, heroic astronauts, exciting discoveries, significant artifacts, and more!

✓ Use this book to see stunning photos from the space age!

✓ I suppose you could bring this book along as very expensive ballast during a space mission, but any reimbursement requests will not be honored!

How Not to Use This Book

✗ Do not use this book as the final word on whether a particular artifact remains on display. Displayed artifacts and exhibits will change over time. Thus, if you're dead set on seeing a particular item, check first with the museum to see if it is still on display!

✗ Do not use this book to design a rocket ship and blast off into space. We're flattered by the consideration, but there are much better books on this subject!

✗ Do not use this book to block cosmic rays during a space mission. Yes, it's a thick book, but it's made of paper, not lead!

Disclaimer

Exhibits and displayed artifacts, like space-flown vehicles, will change over time. Museum displays often include a mix of owned items, which are part of the museum's collection, and borrowed items, on loan from other institutions, including the Smithsonian and NASA. Important to note is that both borrowed and owned items can go on and off display, at times, for various reasons, such as gallery renovations, artifact recalls, or loans to other museums for temporary exhibitions. The listings in this book are representative of what was on display during my research. But they also give an indication of those museums' interest in space. In some cases, when one space item goes off display, the museum may try to replace it with another space-related item. If you hope to see a certain historic item, before you go, check with the museum to confirm it remains on display.

How This Book Is Organized

This book primarily includes two types of chapters: sites and stories. **Site** chapters describe over sixty adventures deemed to have significant relevance to the space age. Plus, within the site chapters, there are **orbital neighbors**, which describe over forty additional space-related highlights that may appeal to readers.

Sites are organized into eight geographic regions across the lower forty-eight U.S. states: Washington DC and Virginia, New York City and the Northeast, Cape Canaveral in Florida, the South and the Gulf Coast, the Midwest, the Great Plains, the Southwest, and the West Coast.

For each site, there is a comprehensive guidebook listing with relevant details to help you plan your trip. For each orbital neighbor, there is a short summary of the highlight with a website address to learn more.

The remaining chapters are **true stories** of the space age: short, compact episodes organized into twenty-four parts and presented mostly chronologically. These stories cover major events and forgotten exploits from the earliest rockets to the modern era. You can read them straight through for a compact overview of the space age or you can jump around to those episodes that interest you the most.

- **Celestial Overview**: Each of the eight geographic regions opens with an overview of its must-see adventures and some notes about the location.
- **Debriefing**: Each listing opens with a brief summary of the adventure. If a museum explores topics other than space travel, such as aviation, there may be a few details about those other highlights.

- **The STARS:** This part of the listing discusses the space age highlights. Listed items may include authentic space-flown and unflown vehicles and artifacts, full-scale replicas, space-related exhibits, displays, and experiences.
- **EVAs:** This usually draws attention to other activities that are either part of the adventure or optional additions. These might include guided tours, self-guided walking tours, films, star shows, simulators, rides, and more.
- **Countdown:** Here you'll find details about when the attraction is open.
- **Mission Budget:** This section offers the estimated cost of the adventure at time of research. This is the price per person/admission ticket unless otherwise noted. ($ = up to $15; $$ = $15–$30; $$$ = $30–$45; $$$$ = $45+)
- **Flight Team:** This includes a phone number and website.
- **Coordinates:** Here you get GPS coordinates and a physical address.
- **Flight Plan:** This includes more information about the location, including any unique details about visiting, parking, etc.
- **Mission Parameters:** If there are any additional details worth sharing about the adventure, they go here. This might include special situations, hazards, tips, etc.
- **Rations:** If the adventure or museum has an on-site restaurant, that is noted here.
- **Orbital Neighbors:** These attractions typically have some relevance to space, or they are smaller facilities, or they're located farther away from most regional attractions. Examples include large museums with a single space-related exhibit, or a random rocket display, or a remote observatory, etc.

Must-See Sites

Each of the space age adventures and orbital neighbors in this book has something relevant to offer, but here are some suggested must-see sites to put on your mission radar. These are not ranked but presented in the order they appear in this book. You'll want to read each listing to decide which sites appeal to you the most. While long-distance road trips could be planned within each region in the book, there are also some sites spread across different regions that would combine well into custom itineraries.

SPACE MUSEUMS

- Smithsonian National Air and Space Museum, Washington DC
- Smithsonian Stephen F. Udvar-Hazy Center, Chantilly VA
- Kennedy Space Center Visitor Complex, Merritt Island FL
- U.S. Space and Rocket Center, Huntsville AL
- Space Center Houston, Houston TX
- Great Lakes Science Center and NASA Glenn Visitor Center, Cleveland OH
- Stafford Air and Space Museum, Weatherford OK
- Cosmosphere Space Museum, Hutchison KS
- California Science Center, Los Angeles CA

SPACE-FLOWN ARTIFACTS

- space shuttle (Smithsonian Stephen F. Udvar-Hazy Center, Chantilly VA; Kennedy Space Center Visitor Complex, Merritt Island FL; California Science Center, Los Angeles CA)
- SpaceX Falcon 9 booster (Kennedy Space Center Visitor Complex, Merritt Island FL; Space Center Houston, Houston TX)
- Mercury capsule (numerous sites)
- Gemini capsule (numerous sites)
- Apollo capsule (numerous sites)
- Moon rock (numerous sites)

UNFLOWN ARTIFACTS AND REPLICAS

- complete Saturn V (Kennedy Space Center Visitor Complex, Merritt Island FL; U.S. Space and Rocket Center, Huntsville AL; Space Center Houston, Houston TX)
- *Skylab* (Smithsonian National Air and Space Museum, Washington DC; U.S. Space and Rocket Center, Huntsville AL)
- lunar module (numerous sites)
- lunar rover (numerous sites)
- Apollo service module/CSM (numerous sites)
- V-2 rocket (numerous sites)

4. A space shuttle solid rocket booster on display outside the Pima Air and Space Museum in Tucson AZ. Mike Bezemek.

UNIQUE ADVENTURES

- Spaceport America, Truth or Consequences NM
- Space and Science Fiction Sites, Southern Utah
- Meteor Crater, Winslow AZ
- Astronaut Training Sites, Northern Arizona
- Nevada National Security Site, Las Vegas NV
- Science Fiction Film Sites, CA

LAUNCH VIEWING

- NASA Wallops Flight Facility Visitor Center, Wallops Island VA
- Kennedy Space Center Visitor Complex, Merritt Island FL
- SpaceX Starbase Launch Viewing, South Padre Island TX

MULTIREGION ROAD TRIPS

- EAST COAST: Choose from the three sites in New York City (Intrepid Sea, Air and Space Museum, American Museum of Natural History, and Cradle of Aviation Museum), plus all five sites in Washington DC and Virginia, plus all three sites near Cape Canaveral FL.
- SUNBELT: Choose from the three sites near Cape Canaveral FL, plus all eight sites in the South and on the Gulf Coast, plus all seventeen sites in the Southwest, and the four southernmost sites on the West Coast (San Diego Air and Space Museum, California Science Center, Griffith Observatory, and More Space Sites, Southern CA)
- HEARTLAND: Choose from all sites in the Great Plains and Midwest
- FAR NORTH: Choose from the three northernmost sites in the Northeast (McAuliffe-Shepherd Discovery Center, Moonshot Museum, Carnegie Science Center), plus six sites in the upper Midwest (Great Lakes Science Center and NASA Glenn Visitor Center, International Women's Air and Space Museum, National Museum of the U.S. Air Force, Air Zoo Aerospace and Science Experience, Museum of Science and Industry, Adler Planetarium), plus two sites from the Great Plains (Strategic Air Command and Aerospace Museum, South Dakota Air and Space Museum), and the northernmost site from the West Coast (Museum of Flight).

SPACE AGE ADVENTURES

The First Rockets

Fire Arrows and Rocket Chairs

No one knows for certain when humans fired the first rocket into the sky. Some suggest the precursor was a wooden bird built in ancient Greece by the philosopher Archytas around 400 BCE. According to the story, written five centuries later by a Roman author, a burst of escaping steam propelled this wooden bird along a suspended wire for several hundred feet.

A more recognizable early rocket was probably invented in China, sometime during the Song Dynasty, around the eleventh or twelfth century. During warfare, bamboo tubes were packed with black powder—also invented in China—and attached to arrows and lances, propelling these fiery projectiles toward the enemy.

In 1232 the Chinese used these fire arrows against the Mongols during the siege of Kaifeng. After a year, Ögedei Khan's army had taken the walled city. Soon the Mongols were using their own primitive rockets as they battled their way across Eastern Europe and the Middle East. As their conquest spread, so did the use of black powder and primitive rocketry.

Some historians suggest the Mongols launched rockets in 1242 at the Battle of Mohi, on the Sajó River in Hungary. Other accounts say the Mongols used rockets during the Siege of Baghdad in 1258. Thirty years later, Arab forces may have attacked the Spanish city of Valencia with rockets. In 1379 an Italian historian named Muratori coined the word *rochetta*, or rocket in English. And the next year such rochettas supposedly flew during a naval battle at Chioggia, near Venice. By 1405 rockets were depicted in an illustrated military manual by German engineer Konrad Keiser.

Primitive rockets continued to be periodically used in warfare during the following centuries, but they typically suffered from limited range and accuracy. Beginning in the fifteenth century, rockets were increasingly supplanted by more powerful technologies like firearms, artillery, and cannons.

An oft-repeated legend from this time, set during the Ming Dynasty in the fifteenth century, tells the story of the world's first attempted astronaut. Supposedly, a Chinese official named Wan Hu built a flying chair with a pair of kites and forty-seven rockets. After his assistants lit the fuses, there was a massive explosion. When the smoke cleared, Wan Hu and his rocket chair were gone.

Epicycles, Heliocentrism, and Laws of Motion

During this same period, generally called the Renaissance in Europe, vast innovations and inventions swept through the "Old World." Common thinking about astronomy and the universe began to evolve. The belief that the earth was not flat but a sphere had been gaining acceptance for centuries, reinforced by the expedition of Portuguese explorer Magellan, which circumnavigated the globe in 1519.

However, for over 1,500 years, Roman and other European scholars continued to believe in a geocentric model of the universe. After all, they stood on a spherical Earth and, when they looked up, the objects in the sky—the Sun, Moon, planets, and stars—all appeared to revolve around them.

Described by the Roman astronomer Claudius Ptolemy in the second century CE, this geocentric model became known as the Ptolemaic system. Ptolemy claimed the universe was divided into a series of nine crystalline spheres in which the features of the observable sky were embedded. The outermost sphere held the heavens. The next held the stars. Beneath these two, the remaining seven spheres rotated in the opposite direction. One sphere each for Saturn, Jupiter, Mars, the Sun, Venus, Mercury, and the Moon—in that descending order.

But for the Ptolemaic model to match the observations of planetary movements, which defied the system by sometimes appearing to move in retrograde, or backward, relative to their normal paths, an even more complex feature was required. The solution was epicycles, a concept where the planets were mounted on smaller circles that rotated within the crystalline spheres. When the pesky planets continued to defy the system, the answer was usually the same. Add more epicycles! The result was that by the fifteenth century, a map of the solar system's Ptolemaic orbital paths looked like a harmonograph—a complex geometric drawing created by looping pendulums.

5. A telescope on display at the New Mexico Museum of Space History in Alamogordo NM. Mike Bezemek.

There had to be a better way to predict the planets' movements. In the early sixteenth century, a Polish mathematician named Nicolas Copernicus took a look at the problem. He gradually came to the conclusion that the earth and planets orbited the Sun, and only the Moon orbited the earth. Copernicus wrote up this theory in a book, *On the Revolutions of the Heavenly Spheres*. But fearing the wrath of the Roman Catholic Church, one of the biggest proponents of geocentricism, he waited decades to publish it. The book came out in 1543, the same year Copernicus died. And for about sixty years, all but a few astronomers mostly ignored it.

Then, in 1609, an Italian math professor and astronomer named Galileo Galilei learned about the invention of the telescope in the Netherlands. Studying the patent design, Galileo began building a series of increasingly powerful telescopes. In November, he trained one of these devices on the Moon. Until that time, the prevailing theory had been that the Moon was perfectly smooth and its face reflected an image of the earth. But with his telescope, Galileo observed craters, mountains, and maria, dark plains once thought to be lunar oceans, like what would soon be named the Sea of Tranquility.

A year later, Galileo turned an even more powerful telescope toward Jupiter, spotting three stars surrounding the planet, all lying in a straight line. Over subsequent nights, he observed these stars circling the massive planet. A few days later, he spotted a fourth. Galileo had discovered the four largest moons of Jupiter. They were later named Io, Europa, Ganymede, and Callisto—collectively, they would be called the Galilean satellites. This discovery, among others made by Galileo and his contemporaries, soon revolutionized astronomy.

Galileo began promoting the Copernican system of heliocentricity and, for his efforts, the Roman Catholic Church subjected him to the inquisition. The judgment was heresy, and the penalty was that Galileo must abandon heliocentrism. He kept a low profile for about a decade, but he eventually resumed his advocacy. A second inquisition resulted in house arrest for the remainder of his life. After his death, Galileo's theories gradually took hold of popular thought, and he became known as the father of modern science.

Born a year after Galileo's death, Sir Isaac Newton was an English mathematician who established key theories about classical physics and gravity. Newton famously observed an apple falling from a tree, which led him to create his theory of gravitation. His three laws of motion include the law of inertia, which states a body in motion will continue moving until acted upon by a force. The second law, force equals mass times acceleration ($f=ma$), describes an object's momentum. And the third law states that for every action, such as a force being applied to an object, there is an equal and opposite reaction. By the time Newton died, in the mid-eighteenth century, his laws of motion had helped establish heliocentrism as the dominant model of the solar system, paving the way for the coming space age.

From Rockets' Red Glare to a New Genre

During the late eighteenth century, the use of rockets as weapons saw a revival. In India, the Mysorean Army invented the iron-cased rocket, using them in battle against the British East India Company. These iron rockets were around a foot long and could travel about half a mile. The British adopted the technology, and by the early nineteenth-century British inventor Sir William Congreve had created increasingly larger versions for the British Army, mostly two to four feet in length, though some were larger.

While the artillery and cannons of the time were still more powerful, rockets had the advantage of being easily deployed in rough terrain or from small boats. The British successfully used them in Europe during the Napoleonic Wars and in the United States during the War of 1812. At the Battle of Baltimore, Francis Scott Key watched the British launch rockets at Fort McHenry. After the Americans won the battle and hoisted the U.S. flag, Key wrote "The Star-Spangled Banner," including the famous line about "the rockets' red glare."

Another major development of the nineteenth century was the Industrial Revolution, as hand production methods were replaced by mechanical manufacturing in factories. Along with these technological innovations came a burgeoning new genre of literature: science fiction. In 1865 French author Jules Verne published *From the Earth to the Moon*. Set in the United States after the Civil War, the novel is about the president of the Baltimore Gun Club, who builds a giant cannon capable of shooting a projectile to the Moon. After the cannon is built in Tampa, Florida, three astronauts board the projectile, and it is successfully launched. The story ends there, but five years later Verne published a sequel titled *Around the Moon*.

The astronauts' voyage takes five days to reach the Moon. Along the way, they dodge an incoming asteroid, which gravitationally alters their course and causes them to orbit the Moon instead of hitting it. They deal with intoxicating gases, determine the Moon is barren, and try unsuccessfully to use rockets to land. Eventually, they plummet back to Earth and land in the ocean, where the projectile floats until they're rescued by the U.S. Navy. Even to this day, observers marvel at how much Verne got right about traveling to the Moon.

Sometimes called the father of science fiction, Verne's work inspired countless writers. In 1869 an American author named Edward Everett Hale published "The Brick Moon" in the *Atlantic*. The story is about the first space station, a 200-foot sphere made of bricks, which launches accidentally with thirty-seven people aboard, who manage to survive the vacuum of space.

As the twentieth century approached, author H. G. Wells offered his own vision with *War of the Worlds*, published in 1898. The story imagines resource-depleted Martians launching a devastating attack on London, only to eventually succumb to Earth-borne pathogens. Published three years later,

Wells's *First Men in the Moon* describes an anti-gravity substance called cavorite that propels a spacecraft to the Moon. There, the protagonists discover a civilization of insect-like aliens. While their primary goal was to entertain, authors like Verne and Wells had created a new genre that would soon inspire an entirely new generation of readers: the first rocket scientists.

Birth of the Space Rocket

Tsiolkovsky's Theories

One of the earliest rocket scientists to be inspired by Jules Verne was Konstantin Tsiolkovsky. Born in western Russia in 1857, from an early age he was fascinated with balloons, airships, and space travel. Becoming a math teacher, Tsiolkovsky worked independently on space theories while living in a log house outside of Kaluga, southwest of Moscow.

While he never built a rocket, he spent decades, near the turn of the twentieth century, creating model rockets and producing theories about space travel. One such theory was a mathematical formula for how a rocket's speed increases relative to exhaust velocity and changing mass as propellant is expended, later called the Tsiolkovsky Equation.

His ideas were mostly ignored, or called impractical, throughout much of his lifetime. Yet many of his theories would prove remarkably prescient. He studied various types of propellant, concluding that liquid fuels and oxidizers, including hydrogen and oxygen, offered the best explosive potential for space travel. He also envisioned that reaching orbit would require rocket trains, later called multistage rockets, in which lower stages detach as their fuel is depleted.

Beginning in 1917 widespread dissatisfaction with the Tsarist monarchy led to the Russian Revolution. In 1921 the new Soviet state recognized Tsiolkovsky for his accomplishments, and he received a lifetime pension to pursue further rocket theories. When he died in 1935, he left all his papers to the government of the Soviet Union.

Robert Goddard: The Moon Man

In 1899 a seventeen-year-old named Robert Goddard climbed a cherry tree behind the family barn in Worcester, Massachusetts. Looking up at the sky, he thought how wonderful it would be to build a device to ascend to Mars.

6. Robert Goddard stands next to his invention, the world's first liquid-fueled rocket, launched in 1926. Esther C. Goddard.

In recent years, he'd been reading science fiction like *War of the Worlds* by H. G. Wells, *Edison's Conquest of Mars* by Garrett P. Serviss, and the books of Jules Verne. When he climbed back down to the ground, he said he'd found his life's purpose.

Within fifteen years, Goddard had become a physics professor. At first, he focused his research on solid-fueled rockets, but soon he switched to liquid fuels, believing they held greater potential for thrust. In 1920 the Smithsonian published his book called *A Method of Reaching Extreme Altitudes*, which introduced his most radical ideas to the world, including rocketing a man to the Moon.

Newspapers of the day mostly mocked him as the "Moon man." A *New York Times* editorial scoffed at the idea that rockets could work in the vacuum of space, given that everyone with half a brain knew a rocket required air to push against—a pervasive misconception at the time. Goddard, the editorial claimed, "seems to lack the knowledge ladled out daily in high schools."

The world's first liquid-fueled rocket launch occurred on March 16, 1926. Standing in a field on his aunt's farm in Worcester, Goddard's contraption looked like a children's jungle gym. The support structure had four diagonal bars, placed upright in the middle. The rocket barely resembled Tsiolkovsky's models or what was to come. For one thing, the combustion chamber and nozzle were near the top, firing down onto a protective cap covering the fuel and oxidizer tanks, with fuel pipes linking the two units.

Upon ignition, a flame shot out from the nozzle. In two and a half seconds, the rocket rose forty-one feet, curved over to the side, and crashed 184 feet away in a cabbage patch. It was the rocketry equivalent of the Wright brothers' 1903 flight at Kitty Hawk. Goddard informed the Smithsonian Institution, his benefactor, of the success but kept the story secret from the press.

He spent the next few years developing the design, including moving the engine to the bottom of the rocket. But most tests met with failure, including a 35-pound rocket that crashed shortly after takeoff. The fireball attracted the attention of area residents, who feared a plane crash, and soon a pair of ambulances arrived. "Moon rocket misses target by 238,799½ miles," mocked one headline in a Massachusetts newspaper.

The news spread quickly, and the Massachusetts fire marshal banned all future flights. But not all reactions were negative. Aviator Charles Lindbergh took an interest in Goddard and arranged for funding from millionaire Harold Guggenheim, which allowed Goddard and his wife to relocate to a ranch in isolated Roswell, New Mexico. By the end of 1930, Goddard and his assistants had launched a rocket 2,000 feet high at a speed of 500 miles per hour, landing it safely with a parachute. Despite the addition of exhaust vanes and gyroscopes, other tests failed.

Funding challenges related to the Great Depression forced Goddard to pause his work for several years. But by the mid-1930s, he got another chance.

Incrementally, his improved rockets flew a mile high, and later a mile and a half. In 1936 the secretive Goddard, who preferred to work alone with only a few assistants, finally acknowledged the first liquid-fueled rocket flight ten years earlier. Meanwhile, the prospect of war in Europe was growing. In the early 1940s, Goddard went to work for the U.S. military on experimental missile designs, and he died during the final weeks of World War II.

German Rocket Clubs and Wernher von Braun

In France, during the 1910s, aircraft designer Robert Esnault-Pelterie—inventor of the aileron and control stick—had shifted his attention to rocketry. He lectured widely on the topic of space travel and coined the word *astronautics* in his comprehensive 1930 book on the subject. In the late 1920s, he established an international astronautics award. He encouraged Goddard to apply, but the American scientist declined, preferring to keep his work private. Instead, the inaugural award went to a German schoolteacher from Transylvania named Hermann Oberth, who was influential in German rocketry circles.

Inspired by the works of Jules Verne, Oberth had used a telescope at a young age to peer at the Moon and stars. While serving in an ambulance unit in World War I, he conducted experiments into the effects of weightlessness using water tanks and sedatives. He also designed a missile he hoped would win the war for Germany, but his plans were rejected by authorities.

After the war, he corresponded with Robert Goddard and pursued a PhD in Heidelberg, but his thesis on rocketry was rejected for being too theoretical. Undeterred, in 1923 Oberth published his thesis as a book, *The Rocket into Interplanetary Space*, which attracted attention among the growing numbers of German rocketry enthusiasts.

Silent science fiction films were widely popular during the cultural heyday of Germany's Weimar Republic. In the late 1920s, director Fritz Lang followed up his dystopian epic *Metropolis* with *Frau im Mond*, or *Woman in the Moon*. This early science fiction film depicted a rocket launch and landing on the Moon. Lang hired Oberth as a consultant, tasking him with designing models for the film and launching a rocket at the premier, but an explosion during testing derailed the project.

Around this time, Oberth was involved with several German rocket clubs, including the Society for Space Travel. One of the club members' first projects was the creation of a rocket-powered race car for manufacturer Fritz von Opel. Other parts of the project involved attaching solid-fueled rockets to planes and train cars. These high-speed spectacles attracted worldwide attention to rocketry. In 1931 the club launched Germany's first liquid-fueled rocket. Burning liquid methane and oxygen, it reached heights of 1,600 feet. In the following months, further launches roughly doubled the peak altitude.

One of the club's most enthusiastic members was a young engineer named Wernher von Braun. Born in 1912 into a noble German family in western Prussia, von Braun grew up mostly in Berlin, where the family moved when he was three. Like other space enthusiasts of the time, he was inspired by the fiction of Jules Verne and H. G. Wells, but also the more practical writings of Hermann Oberth. A fan of the Opel rocket program, when von Braun was a young teenager, he allegedly strapped fireworks to a wagon and piloted this makeshift rocket cart through the streets of Berlin, nearly killing himself in the process.

Von Braun joined the Society of Space Travel several years later. By the time they were launching rockets, he was only twenty. After early successes, the club needed funding. In 1932 they invited German army captain Walter Dornberger to watch a launch. The rocket failed, but Dornberger saw the potential and offered the group a development contract. The requirement was they work in secret on rocket weapons, which hadn't been explicitly banned by the Treaty of Versailles after World War I. The eager von Braun was in favor of accepting military funding, seeing it as the only way to continue the expensive pursuit of rocketry. However, the majority of the club was opposed to military involvement. After a fierce debate, the contract was declined. But Dornberger had taken notice of von Braun, and later that year, he approached the young engineer with a job offer.

3 The V-2

The Aggregats

Wernher von Braun went to work for the German army at Kummersdorf, south of Berlin, in December 1932. Working under Dornberger, von Braun discovered the army had already spent three years researching rockets, though they hadn't gotten very far. The twenty-year-old engineer was soon developing the Aggregat series, an eventual plan for twelve ballistic missiles of increasing size and complexity.

Two months later, the leader of the militant Nazi Party, Adolf Hitler, was appointed chancellor of Germany amid economic unrest related to the Great Depression. A month later, the Reichstag—the German Parliament building in Berlin—was set on fire. Though a Dutch Communist was immediately blamed, some Germans held suspicions the Nazis were behind the fire. Within days, the Nazis had engineered the suspension of civil liberties, including freedom of speech and the press, thus squashing all dissent. Under Hitler's expanding dictatorial powers, Germany quickly increased the secret rebuilding of their armed forces, including the development of new weapons.

The first rocket von Braun built was the A-1. This gyroscope-stabilized missile was about five feet tall and weighed 330 pounds, fueled by an alcohol-water mixture and liquid oxygen. Ready by late 1933, the A-1 exploded during testing, so von Braun proceeded to the next iteration. Two slightly longer A-2s were built in 1934, with each flying over a mile high. Given this success, a new testing ground was needed. Von Braun's mother suggested the isolated peninsula of Peenemünde on the Baltic Sea coast.

Hitler Rises

Meanwhile, Hitler's power was significantly growing. After the death of President Paul von Hindenburg in 1934, Hitler merged the offices of chancellor and president, assuming the title of führer. The following year, he publicly denounced the Treaty of Versailles, revealed the existence of the Luftwaffe air force, and reinstituted a military draft. The next spring, Hitler's troops entered the Rhineland, which since the end of World War I had been a mandatory demilitarized zone along Germany's western border. When neither France nor Britain intervened, Hitler was emboldened.

Civil war broke out in Spain in 1936, and Hitler sent military units to aid General Franco's fascist Nationalists. In 1938 Germany annexed Austria and parts of Czechoslovakia, and the following year it invaded the remainder of the latter country. In secret, Hitler and the Soviet Union's dictator, Joseph Stalin, signed a nonaggression pact, essentially dividing up parts of Europe into German and Soviet spheres of influence. As a second world war seemed increasingly likely, funding for German missiles had been increasing. Dornberger gave von Braun the specifications for the final weapon, the A-4, which was to be many times bigger than the early Aggregats.

At Peenemünde, by 1937, much progress had been made on an intermediate design, the A-3. The shape of this 22-foot rocket was based on the 8mm infantry bullet. It weighed 1,650 pounds and carried an advanced guidance controller. Still, most A-3s crashed in similar fashion. The parachutes deployed early and the engines failed, with the rockets arcing over and crashing into the sea. Data pointed to design flaws, and thus another intermediate test vehicle was needed. The A-5 was roughly the same size but carried an improved guidance system. A half dozen were successfully launched during late 1938 and throughout 1939.

In March of 1939, on von Braun's twenty-seventh birthday, Hitler came to Kummersdorf for a test firing of rocket engines. The führer seemed unimpressed by the roaring exhaust. But he grilled von Braun and Dornberger about its potential as a weapon, and the work was continued as World War II began. First, the Nazis invaded Poland from the west in September 1939. Ten days later, Stalin ordered the Red Army to invade from the east. The campaigns were successful, and the two occupiers split the country between themselves.

Being allied with Poland, Britain and France declared war on Germany, but they did nothing. This initiated an eight-month period of military buildup without fighting known as the Phoney War. While Stalin threatened the Baltic states and invaded Finland in the Winter War, Hitler turned his sights toward western Europe. The Nazis waited until spring 1940, when they used blitzkrieg tactics to shockingly overrun the Low Countries and France in only six weeks.

The Era of Space Travel

As 1940 arrived Peenemünde had grown to over 1,000 personnel. To study the aerodynamics of the A-4, which would fly supersonically, a large wind tunnel was built. By 1941 the A-4 was ready for flight tests. It stood 46 feet tall and weighed over 27,000 pounds, carrying a one-ton warhead of mostly TNT. Essentially an enlarged A-5, the A-4 was also shaped like a bullet with four stabilizing tail fins. The rocket generated 56,000 pounds of thrust, and it was guided by four external rudders and four internal graphite vanes that directed the exhaust.

The first two A-4 rockets to be launched broke the speed of sound but came apart in flight and crashed into the Baltic Sea. But on October 3, 1942, the third A-4 rose successfully to an altitude of 56 miles before impacting the sea 120 miles downrange. It was the world's first ballistic missile and, in a later test, it would become the first manmade object to reach outer space.

"The first day of a new era," commented Dornberger. "That of space travel."

But the immediate concern was completing a weapon that might save the Nazi Party. In July 1942 Dornberger and von Braun brought a film of the successful October launch to show Hitler. This time, seeking any advantage to turn the tide of the war, the führer was impressed. By now, all hopes of a swift Nazi victory had been dashed. The British had repelled the Luftwaffe during the nearly-four-month Battle of Britain, a Nazi bombing campaign that had progressed from military targets to civilians, killing over 23,000 people. After invading the Soviet Union, the Germans were making steady but slow progress against the Red Army. And following the December 1941 Japanese attack on Pearl Harbor, the United States had declared war on the three Axis powers, Japan, Italy, and Germany.

Throughout the war, the Allies conducted a massive bombing campaign over Germany, also progressing from military to increasingly civilian targets. There was a full Moon over the Baltic coast on the night of August 18, 1943. Just past midnight, almost 600 aircraft of the Royal Air Force began bombing Peenemünde. The first target was the residential compound housing the rocket scientists, but a targeting error caused the first bombs to land on the forced labor camp, killing an unknown number of imprisoned workers.

By the time the bombing was complete, much of the base was in ruins. While some German technicians were killed, all but one of the critical A-4 scientists made it to safety. To deter future bombings, the base was left partially destroyed while design work continued on the premises. In the aftermath, Hitler ordered that production of the A-4 be moved to the Mittelwerk tunnel complex in the Harz Mountains, where slave laborers—inmates of Nazi concentration camps—were forced to build thousands of missiles in deplorable conditions.

From Vengeance to Surrender

In 1944 the A-4 rocket was renamed the Vergeltungswaffe 2, or Vengeance Weapon 2. Intended to retaliate for the massive devastation inflicted by the Allies across Germany, the V-2 would join another such missile. The V-1 Flying Bomb was an early cruise missile, propelled by a pulsejet combustion engine. Simultaneously developed at Peenemünde, the V-1 was also known as the buzz bomb for the humming sound it made upon approach, before going silent during an unpowered final descent.

By now, a thirty-two-year-old von Braun was fully entrenched in the Nazi war machine. As a condition for completing his rocket work, he'd previously joined the Nazi Party and later the SS, a paramilitary unit that enforced Nazi control, including concentration camps. In March, an SS spy overheard von Braun and some colleagues discussing the fact that the war wasn't going well for Germany, expressing regret that they weren't working on space travel endeavors. The SS arrested von Braun, and he spent two weeks in jail before Dornberger secured his release.

On June 6 the Allies landed at Normandy in the D-day invasion, beginning a slow campaign of liberation that would reach Paris by late August and continue onward into Germany. Soon after D-day, the Nazis were deploying V-1 mobile launch units to the western front, which were joined by V-2 units in September. The main targets were civilian areas in London and in Antwerp, Belgium. During the following months, nearly 3,200 V-2s, each with a maximum range of about 200 miles, were launched at these Allied cities, killing around 5,000 people. As the Germans retreated, the V-2s were forced out of range, and the Allies unleashed their own vengeance—a fire-bombing campaign that leveled German cities like Dresden and killed tens of thousands.

By January of 1945 the Soviet army was about 100 miles east of Peenemünde. Orders sent the scientists south to Mittelwerk, where von Braun had thousands of V-2 documents hidden in an abandoned mine. As the Allies closed in from the west, the German scientists moved farther south to the Bavarian Alps, where they hid in small mountain villages.

In Berlin, on April 30, Hitler shot himself in his bunker. After two weeks of fierce urban fighting, Soviet forces raised their flag over the German Reichstag. That same day, hundreds of miles south in Bavaria, von Braun's brother Magnus rode a bicycle to meet an American patrol. "My name is Magnus von Braun," he said. "My brother invented the V-2. We want to surrender."

Sites in Washington DC and Virginia

Five Adventures and Three Orbital Neighbors

Celestial overview: For many enthusiasts, the exceptional Smithsonian National Air and Space Museum on the National Mall in Washington DC is *the* pilgrimage site for historic space artifacts. For that reason, it's the first adventure in this book. But keep in mind that the surrounding region boasts at least four additional sites, each offering something a little different. At the top of that list is the Smithsonian's Udvar-Hazy Center. Located in an eastern suburb of DC, this is the Air and Space Museum's massive annex where many of the collection's top spacecraft and aircraft are on display, including a space shuttle and much more!

Just east of DC, the NASA Goddard Visitor Center is a small facility with artifacts and exhibits that will appeal to diehard space enthusiasts. Similarly, to the south on the Delmarva Peninsula, the NASA Wallops Flight Facility Visitor Center is located at an active rocket test site best known for launch viewing (advance planning required). From there, it's not far to the final regional adventure, the Virginia Air and Space Science Center in Norfolk. The center, which also serves as the NASA Langley Visitor Center, houses impressive displays of spacecraft, replicas, and exhibits. Put it all together and you could plan an exciting space-focused vacation around these five sites, perhaps tacking on additional stops at orbital neighbors or places in New York City and the Northeast region as well.

Smithsonian National Air and Space Museum, Washington DC

Debriefing: The Smithsonian National Air and Space Museum in Washington DC is one of the nation's most visited museums. The expansive galleries include some of the most iconic flown artifacts of the aviation and space ages. During research for this book, the museum was completing a major renovation to its galleries, and precise information about the final displays was subject to change. While most of the highlights discussed below should remain on display, some may shift to the Udvar-Hazy Center and vice versa. Check the museum website or call for updates.

One of the most famous aviation artifacts is the original 1903 *Wright Flyer*, which made the historic Wright brothers flight from Kitty Hawk, surrounded by a comprehensive exhibition about the birth of the aerial age. An early-flight gallery focuses on developments from 1903 until the outbreak of World War I in 1914, including period aircraft. Two other famous historic artifacts are the *Spirit of St. Louis*, flown by Charles Lindbergh on the first solo nonstop transatlantic flight, and the *Glamorous Glennis*, the Bell X-1 rocket-powered aircraft piloted by Chuck Yeager to break the sound barrier in 1947.

The Jay I. Kislak World War II in the Air gallery will focus on the wartime revolution in flight, with displayed fighters including the P-51D Mustang, the FM-1 Wildcat, and the Messerschmitt Bf 109G. The Thomas W. Haas We All Fly gallery will focus on general aviation and aircraft. The America by Air

7. For many years, one of the highlights of the Smithsonian collection has been the Apollo-Soyuz display. Mike Bezemek.

gallery will focus on the history of air transportation in the United States, including the roles of technology and the federal government.

The STARS: The museum is home to many beloved space artifacts, both large and small. One favorite is the backup-flight-worthy *Skylab* Orbital Workshop, the first U.S. space station modified in the 1970s from an Apollo program Saturn IV-B upper stage. Visitors can walk through the *Skylab* living quarters.

One of the biggest and most impressive displays is a unified recreation of the Apollo-Soyuz Test Project. The Apollo CSM (command service module) and Soyuz spacecraft are full-size test vehicles, while the docking module is an unflown backup. This joint U.S.-U.S.S.R. mission docked in orbit in 1975, signaling the end of the Space Race.

The Destination Moon gallery will feature a variety of space-flown artifacts, including Alan Shepard's spacesuit from the first U.S. suborbital flight and a capsule from the first orbital mission, *Friendship 7*. The capsule from the marathon fourteen-day *Gemini 7* mission in 1965. The historic *Apollo 11* command module Columbia. The spacesuit Neil Armstrong wore on the lunar surface. Full-size testing units include an Apollo lunar module, a lunar rover, and a Surveyor lunar lander, five of which were successfully soft-landed in the 1960s.

Other missiles and rockets include an authentic V-2 captured by U.S. forces after World War II, a Minuteman III, a Standard Missile-3, and others. A full-scale testing mock-up of the Hubble Space Telescope has been refurbished to match the appearance of the real Hubble Space Telescope in orbit. Alexei Leonov's training spacesuit and Volga inflatable airlock were used before the first human spacewalk on *Voskhod 2*.

The Kenneth C. Griffin Exploring the Planets Gallery will focus on the historic events and scientific discoveries of planetary exploration. Highlights will include a full-size testing mock-up of a Voyager spacecraft, two of which launched in 1977 and entered interstellar space in the 2010s. The *Marie Curie* rover, the flightworthy backup for *Sojourner*, the first successful rover on Mars when it landed in 1997. A full-scale model of the *Curiosity* rover, which landed on Mars in 2012 to search for signs of past habitability. A photomosaic globe of Mars using satellite imagery from *Mariner 9*.

Elsewhere, a pair of exhibitions explores links among air, space, and culture. The Nation of Speed gallery will focus on how the pursuit of speed has shaped the American identity across land, sea, air, and space—including NASA's hypersonic rocket drone, the X-43. One World Connected examines how aviation and spaceflight, including satellites, changed everyday life and evolved our understanding of Earth.

EVAs: The museum has an IMAX theater offering a variety of relevant films, plus there are simulators and virtual reality experiences. Paid tickets are available online or in person. Science demonstrations are often held throughout the museum, and guided tours led by experts are available.

Countdown: Open daily 10:00 a.m.–5:30 p.m.

Mission budget: Free

Flight team: 202-633-2214 | https://airandspace.si.edu

Coordinates: 38.8884, -77.0199 | 655 Jefferson Drive SW, Washington DC 20560

Flight plan: The museum does not offer parking, but there are private parking facilities located nearby. There is a limited amount of metered street parking around the National Mall. Public transportation includes Metrobus and Metrorail, with several stops nearby. Visit the museum website for more information.

Mission parameters: Located on the National Mall, where visitors often walk between museums and the U.S. Capitol Building. Some of these museums have exhibits related to space, including meteorites at the **National Museum of Natural History**. Other excellent museums include the **National Museum of the American Indian**, the **National Museum of African American History and Culture**, the **National Museum of American History**, and the **Arts & Industries Building**, which in 2022 held an exhibition titled "Futures" that included several space-related exhibits.

Rations: During busy days like weekends, holidays, and summer vacation, food trucks often line the cross streets around the National Mall.

Orbital neighbors: Consider a stop at **The Castle**, the original Smithsonian Institution Building, located nearby on the National Mall. Exhibits present the story of the Smithsonian Institution with a small display of artifacts from each museum, including a few space items. **Flight team:** www.si.edu /museums/smithsonian-institution-building.

Smithsonian Stephen F. Udvar-Hazy Center, Chantilly VA

Debriefing: The Stephen F. Udvar-Hazy Center, near Washington Dulles International Airport, is the annex of the Smithsonian Air and Space Museum. Inside two massive hangars, the facility displays thousands of air and space artifacts and vehicles. Combined with a visit to the main Smithsonian Air and Space Museum, discussed above, these two facilities are a must-see for space enthusiasts. In addition to an impressive space gallery, most of the Udvar-Hazy Center displays a mix of rare, popular, and famous aircraft. The aviation highlights are plentiful, including an Air France Concorde supersonic airliner, an SR-71 Blackbird, and the *Enola Gay*, the Boeing B-29 Super Fortress that dropped an atomic bomb on Hiroshima during World War II.

The STARS: The James S. McDonnell Space Hangar is where you'll find most of the museum's space artifacts. The biggest highlight is the space shuttle *Discovery*, displayed horizontally in the center of the hangar. The third shuttle to launch into space, *Discovery* was the most accomplished orbiter in the fleet, flying thirty-nine missions from 1984 to 2011.

Surrounding *Discovery*, visitors will find other exhibits and artifacts related to the shuttle program. Hanging nearby, there's a Manned Maneuvering Unit, a propulsion backpack that allowed astronauts to maneuver outside the spacecraft. Next to the orbiter is the Canadarm Remote Manipulator System, a robotic arm used to position payloads and astronauts outside the spacecraft. Elsewhere, the Spacelab Laboratory Module was an orbital research facility flown in the shuttle payload bay.

A series of NASA artifacts focuses on human spaceflight. From Project Mercury, there's the Big Joe capsule from an unmanned test flight and an unflown capsule from the canceled *Mercury-Atlas 10*. One unique artifact is the Gemini TTV-1 Paraglider Capsule, an early test vehicle used to practice landing the capsule like an aircraft on land.

From the Apollo program there's a boilerplate command module, displayed with the actual flotation collar and air bags from *Apollo 11*, that was used as a water egress trainer. Also on display is the *Apollo 11* mobile quarantine facility, the modified Airstream trailer used to transport Armstrong, Aldrin, and Collins upon their return from the Moon.

Other exhibits focus on rocketry and unmanned spacecraft. Among these, there's a replica of the world's first liquid-fueled rocket, launched by Robert Goddard on March 16, 1926. Nearby is Goddard's final rocket, developed in New Mexico and tested in the early 1940s.

Unmanned spacecraft include a tracking and data relay satellite, one of three launched into orbit in the 1980s to provide nearly constant contact with orbiting spacecraft. A set of artifacts relates to the Soviet Union's exploration of the inner solar system: engineering models from the Vega project, which sent two probes to Venus in the mid-1980s, including the main spacecraft bus, the lander, and a balloon with scientific instruments that floated through the Venusian atmosphere. There's also a full-scale prototype of the Mars *Pathfinder* lander and rover, which landed on the Red Planet in 1997. There's plenty more to see at the center, and a full list of the artifacts and exhibits on display can be found on the museum website.

EVAs: An IMAX theater regularly shows 2D and 3D giant-screen films related to air and space topics, plus there are ride simulators and interactive flight simulators (tickets available online or in person at the box office). An observation tower offers a 360-degree view of Washington Dulles International Airport, allowing visitors an opportunity to observe airplane landings and takeoffs. Throughout the museum, there are also regular science demonstrations and experts available to answer questions.

8. The space shuttle *Discovery* on display in the McDonnell Space Hangar at Udvar-Hazy in Virginia. Mike Bezemek.

Countdown: Open daily 10:00 a.m.–5:30 p.m. (except December 25)

Mission budget: $ (museum entry is free but there is a per-vehicle parking fee)

Flight team: 703-572-4118 | https://airandspace.si.edu/udvar-hazy-center

Coordinates: 38.9114, -77.4426 | 14390 Air and Space Museum Parkway, Chantilly VA 20151

Flight plan: The center is a few miles north of I-66 in Chantilly, VA. Depending on traffic, it's a minimum thirty-minute drive from the main Smithsonian Air and Space Museum in DC.

Mission parameters: The Udvar-Hazy Center can be reached by public transportation from points around the DC metro area and the National Mall using a combination of the Washington Metro and bus service. Visit the center website for current information.

Rations: The Shake Shack is the center's on-site restaurant. There are many more restaurants available throughout the surrounding area.

NASA Goddard Visitor Center, Greenbelt MD

Debriefing: Located about six miles northeast of Washington DC, the Goddard Space Flight Center is one of NASA's ten major research laboratories. Named after American rocket pioneer Robert Goddard, the center is focused on space-based observations of the earth, Moon, Sun, and solar system. Projects include the early Explorer satellites, the Hubble Space Telescope, the Lunar Reconnaissance Orbiter, and the recent James Webb Space Telescope. The spaceflight center itself can't be visited by the general public, but the visitor center is a free site open to the public with spacecraft and exhibits related to technologies developed at Goddard.

The STARS: Outside the visitor center, the Goddard Rocket Garden and Astrobiology Walk features space artifacts and discoveries related to the center. A small rocket park includes an early Delta rocket and some smaller missiles and sounding rockets. Nearby, the Goddard Moon Tree is one of several around the country grown from seeds that flew to the Moon during *Apollo 14*.

Inside, Solarium is a digital art installation that visualizes solar flares and eruptions. The Decade of Light is an exhibit exploring space communication and navigation. Globe Hall allows visitors to participate in the educational Globe Program, while Neighborhood Earth lets visitors explore Earth from the point of view of orbiting satellites.

Lunar exhibits include a full-size mechanical prototype of the Lunar Reconnaissance Orbiter, plus an exhibit exploring the Apollo landing sites and a Moon rock brought back by *Apollo 11*. Other exhibits focus on the Hubble and James Webb space telescopes.

EVAs: Group tours of Goddard Space Flight Center may be possible with advance reservation.

Countdown: September–June: Tuesday–Friday 10:00 a.m.–3:00 p.m.; July–August: Saturday and Sunday 12:00–4:00 p.m.

Mission budget: Free

Flight team: 301-286-8981 | www.nasa.gov/centers/goddard/visitor/home /index.html

Coordinates: 38.9927, -76.8467 | 9432 Greenbelt Road, Greenbelt MD 20771

Flight plan: The visitor center is located off ICESat Road. From Route 295, take Greenbelt Road East for about 2.5 miles. Turn left onto ICESat Road and turn left at the first opportunity to reach the visitor center parking lot.

Rations: There's no food at the center, but restaurant options can be found throughout the surrounding area.

Orbital neighbors: Located in Linthicum MD, near Baltimore, the **National Electronics Museum** has a small exhibit related to orbital satellites, which may be of interest. **Flight team**: www.nationalelectronicsmuseum.org.

NASA Wallops Flight Facility Visitor Center, Wallops Island VA

Debriefing: Located on a Virginia island off the east coast of the Delmarva Peninsula, NASA's Wallops Flight Facility is an active rocket launch site and flight test airfield. The facility is managed by the NASA Goddard Space Flight Center in nearby Greenbelt MD. Since being established in 1945, Wallops has seen thousands of launches. These include test flights for Project Mercury using the Little Joe rocket that carried the rhesus monkeys Sam and Miss Sam. Today, Wallops offers a visitor center with exhibits and the opportunity to witness rocket launches.

The **STARS**: Visitor center exhibits focus on aeronautical science and the facility's endeavors, including Wallops history, current space programs, and upcoming missions. The visitor center offers free educational programs and weekly public activities, some of which are linked to astronomical events. There is a rooftop viewing area and ground-level bleachers for watching rockets launching from several coastal pads about seven miles from the visitor center. This rooftop viewing area is one of the few public sites with a clear view of the launch sites. Viewing is free but space is limited and available on a first-come, first-served basis only. Launches of smaller sounding rockets rarely fill to capacity, while larger launches, including the Antares rocket, often fill to capacity about 60–90 minutes after the center opens.

Mission parameters: Watching a rocket launch is one of the most popular activities at Wallops, but it involves a few challenges. Launch schedules frequently change, and dates are not typically released more than two months in advance. Launch days and times may be changed due to weather, sometimes right before launch. Sounding rocket launches are very quick events, and the viewing area rarely fills to capacity. But larger launches are very popular events, with thousands of people often attending. As a result, NASA

9. An Antares rocket launches from Wallops in 2017 on a cargo supply mission to the International Space Station. NASA Wallops Flight Facility/ Patrick Black.

recommends you arrive within a half hour of the visitor center opening on the day of a launch. If the viewing areas are full, it is recommended to have a backup viewing location in mind, with information listed for several sites on the visitor center web page.

Countdown: During early 2022 the visitor center remained closed due to COVID-19, but it will likely reopen in the future, so check the website for updates.

Mission budget: Free

Flight team: 757-824-1404 | www.nasa.gov/content/nasa-wallops-visitor-center-2

Coordinates: 37.9387, -75.4571 | Building J20, Route 175, Wallops Island VA 23337

Flight plan: The visitor center is located on Route 175, six miles east of US-13.

Rations: Restaurants and stores can be found on nearby **Chincoteague Island**.

Virginia Air and Space Science Center, Hampton VA

Debriefing: The Virginia Air & Space Science Center is also the NASA Langley Visitor Center. The Langley Research Center was the first civilian aeronautical laboratory in the United States, founded under the National Advisory Committee for Aeronautics in 1917. In addition to having many impressive space displays, the science center has over twenty aircraft on display, highlighting 100 years of aviation history, including a replica 1903 *Wright Flyer*, a DC-9 passenger jet, a Huey helicopter, and more.

The STARS: One highlight here is the space-flown Apollo 12 command module, the second mission to land on the Moon in November 1969. There's also the Apollo Lunar Excursion Module Simulator, a manned rocket-powered vehicle that Apollo astronauts used in aerial training on Earth to simulate the handling characteristics of the lunar module.

There's the space-flown unmanned Mercury #14 capsule, which was launched twice on Little Joe rockets to test the launch-escape system. Plus a Gemini test capsule that astronauts used to train for emergency water escapes.

27

Two Mars-related items include a replica of a Viking lander, two of which made history, in 1976, as the first successful spacecraft to land safely on Mars. The other item is a model of the *Opportunity* rover, which landed on Mars in 2004 and operated for a remarkable fourteen years.

Related to NASA's current efforts, there is an Orion PA1 flight test vehicle, which was used for the first in-flight test of the Orion launch abort system in May 2010. Meanwhile, the Space Explorer Gallery offers interactive exhibits related to exploring the solar system, including the efforts at Langley Research Center. The Solarium offers an up-close look at the Sun by journeying inside the solar atmosphere using images recorded by NASA's Solar Dynamics Observatory. The Exploration Station is an interactive 80-inch touchscreen that allows viewers to explore Earth like a NASA scientist using satellite technology and other field data.

EVAs: An IMAX theater shows films related to science and space. Tickets are included with general admission.

Countdown: Monday–Saturday 10:00 a.m.–5:00 p.m.; Sunday 12:00–5:00 p.m.; closed Mondays during the off-season (check website for updates)

Mission budget: $$

Flight team: 757-727-0900 | https://vasc.org/

Coordinates: 37.0241, -76.3442 | 600 Settlers Landing Road, Hampton VA 23669

Flight plan: The science center is located in downtown Hampton. There is a free parking garage across the street.

Rations: The **Cosmic Cafe** is on-site. More restaurant options can be found throughout the area.

Orbital neighbor: Located in the mountains of West Virginia, the **Green Bank Observatory** is a radio telescope facility and research center, home to the world's largest fully steerable radio telescope. Visitors can take a variety of guided tours around the facility, learn about radio astronomy in the on-site science museum, and walk the grounds on a self-guided tour.

Flight team: https://greenbankobservatory.org.

4 The Cold War

An Amerika-Rakete

While von Braun and associates were being interrogated, in spring 1945 the U.S. Army sent a team to the recently captured Harz Mountains with orders to retrieve a hundred V-2 rockets. The area was scheduled to be handed over to the Soviet administration in June, so they had to work fast.

Inside the Mittelwerk tunnels, the team found deplorable work conditions and thousands of rocket parts. Without a list of what was needed, they collected a hundred of every part they could find. They loaded the materials into hundreds of train cars, which were sent to Antwerp and loaded onto Liberty cargo ships for the Atlantic crossing.

Another team retrieved von Braun's documents from the mine shaft. These revealed the full scope of the Aggregat series, including a design for an unbuilt 66-foot-tall Amerika-Rakete. Launched from Europe, this missile would have been piloted on a suicide mission toward targets on the U.S. mainland.

The Chief Designer

On May 5, 1945, the Soviet Red Army reached Peenemünde. The base was mostly deserted, with much of the equipment destroyed by explosives. A few months later, the Soviets sent a team of rocket scientists, which included a newly commissioned colonel named Sergei Korolev.

Born in northern Ukraine in 1907, Korolev developed an interest in aviation at a young age and became a pilot as a teenager. A few years after the end of the Russian Revolution, he moved to Moscow to study engineering. Afterward, he went to work for a government aviation design bureau.

During the early 1930s, the same rocket craze that had gripped Weimar Germany was spreading across the Soviet Union. With a growing interest in space exploration, Korolev joined the Moscow chapter of the Group for Study of Reactive Motion and was eventually named the leader. By 1933 they

10. A 1969 postage stamp commemorating Sergei Korolev. Soviet Union.

had launched their first successful liquid-fueled rocket, which flew to an altitude of 250 feet. When the watchful Soviet government learned of their successes, the group was merged into the Red Army. Korolev was appointed deputy chief, and work continued on missiles and rocket-powered aircraft.

In the mid-1930s Soviet dictator Joseph Stalin became paranoid that rivals were plotting to oust him. In 1937 he initiated the Great Purge. Hundreds of thousands of suspected traitors were murdered or imprisoned, often without evidence. One of those arrested was rocket engine designer Valentin Glushko. During an interrogation, he implicated his colleague. Korolev was accused of purposely delaying missile work, which had fallen far behind von Braun's efforts for Nazi Germany.

Korolev was first sent to a Siberian gulag, where he was forced to work in a gold mine. Conditions were so harsh he lost most of his teeth from malnutrition during the first few months. His request for a retrial was granted, and he was transferred to a labor camp for scientists and engineers, where he worked with his former accuser, Glushko. Throughout World War II, they mostly developed bomber aircraft for the Soviet military.

During the final year of the war, Korolev was released and commissioned as a colonel in the Red Army. When he arrived in war-torn Germany, his task was the recovery of V-2 rockets and information. The Soviets soon discovered that almost all the high-level rocket designers had surrendered to the Americans. Only a few German rocket scientists remained, while the rest were mostly technicians involved with mass production of the missile. Scrounging for remaining parts, the Soviets gathered the hardware to rebuild about a dozen V-2s. Moving the German scientists and captured rocket parts east, Korolev's next goal was to reverse engineer the V-2 and build the Soviets their own set of ballistic missiles.

The Manhattan Project

With the war over in Europe, all attention had shifted to the Pacific. In a series of bloody campaigns, American-led forces defeated the Japanese at Iwo Jima and Okinawa. The Allies demanded total and unconditional surrender, but Japan refused, hoping to hold out for better terms. During these last battles of the war, as the situation looked increasingly dire, the Japanese turned to kamikaze attacks. In suicide missions, fighter pilots crashed their fuel-laden planes into Allied warships. Japan had vowed to fight to the death, and Allied plans for invading their islands included estimates of over a million casualties. An alternative was desperately needed.

While the Nazis had focused efforts on developing the two vengeance missiles, the U.S. had pursued a different type of weapon. Called the Manhattan Project, the development of an atomic bomb was so secretive that only a few dozen of over 100,000 workers were aware of the objective. In the predawn hours of July 16, 1945, a fiery explosion illuminated the mountainous landscape near Socorro, New Mexico. The heat melted the sandy desert floor into glass. The mushroom cloud rose almost eight miles high.

A coded message was sent to the American leadership at the Potsdam Conference in occupied Germany. The U.S. president was now Harry Truman, who had succeeded from the vice presidency after an ailing Roosevelt died in April. The leaders of the United States, Britain, and the U.S.S.R. were negotiating the postwar order in Germany and around the world. Truman mentioned to Stalin something vague about a new super weapon.

Three weeks later, on August 6, a customized B-29 bomber named the

11. A two-stage V-2 modified by the U.S. Army for Project Bumper is launched from Cape Canaveral in 1950. NASA.

Enola Gay dropped an atomic bomb on Hiroshima. In an instant, the central core of the key military city was destroyed. A firestorm engulfed two-thirds of the buildings. As many as 20,000 soldiers died along with 75,000 civilians within hours. After announcing the first use of a nuclear weapon in war, Truman demanded unconditional surrender. Japanese scientists confirmed the bomb, but still no surrender came. Three days later, the United States dropped a second atomic bomb on the industrial city of Nagasaki. In a flash, about 40,000 people died, almost entirely civilians, about half of whom worked in war-related factories. Meanwhile, six hours earlier, the Soviet Union had declared war and invaded Japanese-occupied Manchuria. The combined news swayed the Japanese leaders, and six days later a delegation signed surrender documents on the USS *Missouri*.

Operation Paperclip

By now, the U.S. Army had approved Operation Paperclip, which secretly relocated 120 captured German rocket scientists to the United States. They were given six-month work contracts to build facilities and launch the captured V-2 rockets. In September, von Braun was the first to arrive. The site was White Sands Proving Grounds, in New Mexico, where the Germans sorted through parts to reassemble the V-2s.

The first launch was a failure when a tail fin fell off, but successful launches followed. One V-2 even errantly landed over the border in Juarez, Mexico. Soon they were launching V-2s with scientific payloads up to a hundred miles high. Instruments studied the upper atmosphere and incoming solar radiation. Sending 35mm cameras aloft, scientists obtained the first black-and-white images of the earth from space. Other V-2s were modified by adding the slender WAC-Corporal sounding rocket as an upper stage. One of these set a world record by reaching 250 miles high. While these efforts failed to attract much public interest, the response from scientists was sheer enthusiasm. Studying the earth from above carried tremendous benefits.

The military was more interested in the implications for missile development, recognizing they were now in an arms race with the world's other superpower. Since the end of World War II, tensions with the Soviet Union had mounted. Territorial occupations and agreements had effectively split much of Europe and Asia into zones of Soviet and American influence. Several countries, like Germany, Vietnam, and Korea, had been partitioned into communist and capitalist states. To prevent further spread of communism, Truman adopted a doctrine of containment, using various means short of direct confrontation to check Soviet aspirations.

It was these tensions, and fears of another devastating worldwide war, that fueled a dramatic escalation in weapons development. The Soviet Union detonated their first atomic bomb in northeastern Kazakhstan in 1949. Soon, both superpowers were engaged in a back-and-forth series of ever more powerful nuclear bomb tests. In the beginning of this nuclear arms race, the only delivery method was long-range bombers. But these large planes flew slowly and were at risk of being shot down before reaching their targets. Thus, both parties recognized the importance of developing high-speed ballistic missiles capable of traveling through space to deliver heavy nuclear warheads.

In the United States, all three branches of the military began missile development programs. The air force began work on a large intercontinental ballistic missile called the Atlas. The navy focused on the Viking project, a small sounding rocket that might be launched at sea. And the army pursued shorter-range ballistic missiles at the Redstone Arsenal in Huntsville AL. After launching over sixty captured V-2s, mostly at White Sands, the army learned a great deal about rocketry from the Germans. In 1950 von Braun and his team were transferred to Redstone Arsenal to begin work on a new project for the U.S. Army.

5 Conquering Space

From Redstone to Collier's

Upon arriving in Huntsville in 1950, Wernher von Braun led a team of Germans and Americans developing the U.S. Army's first ballistic missile. The goal was an enlarged derivative of the V-2 based partially on designs from the original Aggregat series. This short-range rocket would burn alcohol and liquid oxygen to deliver a 6,900-pound nuclear warhead about 200 miles.

The result was the needle-shaped Redstone, standing just under 70 feet tall and weighing over 60,000 pounds. Built by Chrysler, and using a single Rocketdyne A-7 engine, it launched with 78,000 pounds of thrust. By 1953 the first test launches were being conducted at Cape Canaveral in Florida.

Outside of work, von Braun enjoyed opportunities to discuss space exploration that came with leaving behind the oppressive Nazi regime for the relative freedom of American society. In 1951 science writer and early space advocate Willy Ley organized the first Symposium on Space Travel at New York's Hayden Planetarium. Born in Germany, Ley had been a member of the same rocket club as von Braun before fleeing to America in 1935. In attendance at the New York symposium was a pair of writers from the popular *Collier's* magazine.

Over the next few years, the magazine published an article series on the potentials for space travel titled *Man Will Conquer Space Soon!* Mostly written by von Braun and other experts like Ley, the articles were creatively illustrated by artist Chesley Bonestell. Over the course of eight issues, von Braun led the way in envisioning detailed plans for the future of space travel. There were also urgent warnings that the United States must move fast to establish space superiority, ahead of their Cold War adversaries in the secretive Soviet Union.

A Methodical Conquest

The first step in von Braun's conquest of space was a baby space station, ready in five to seven years if started immediately. Built into a rocket nose cone, this

12. Artist Chesley Bonestell created this drawing for a "Man on the Moon" article, cowritten by von Braun for *Collier's* magazine in 1952. Chesley Bonestell for *Colliers*.

satellite would be powered by solar panels and carry three rhesus monkeys to test the survivability of acceleration, weightlessness, and unknown space phenomena. Next would come a winged spaceplane, launching vertically as part of a three-stage rocket but landing on a runway like an airplane to be serviced for reuse. The cargo bay of this spaceplane would be used to build America's first permanent space station.

The 250-foot-wide station would be shaped like a wagon wheel, rotating to produce artificial gravity through centrifugal force. The design called for nylon-plastic fabric compartments to be inflated in space and later covered by a metallic outer skin. For defense purposes, both the spaceplane and the space station would be armed with missiles—including nuclear warheads—offering Americans an unprecedented military advantage. Cameras trained downward would provide detailed observations of all points on Earth.

Orbiting near the space station would be an unmanned space telescope. Operating outside the distortions of Earth's atmosphere, this telescope would return stunning images and astronomical observations of the distant universe. And servicing all these orbiting facilities would be spacemen in pressure suits, floating attached to tethers and propelling themselves with handheld rockets.

Expeditions: Moon and Mars

Once the station was ready, the next goal was the Moon within twenty-five years. Von Braun's vision was a large landing party, like a lunar-bound expeditionary force from World War II. Three lunar spacecraft, each 160 feet long, would be assembled in Earth orbit. The spherical crew compartments would carry fifty explorers to the Moon in five days. After jettisoning its external departure fuel tanks, the craft would become an awkward lander with sharp angles and extendable legs for landing vertically on the lunar surface.

After the crew emerged in spacesuits, they would board tank-like tractors to find a suitable deep crevice offering shelter from cosmic rays and meteorites. From above, the crew would lower prefabricated huts to serve as their living quarters and laboratories for a canyon-floor Moon base. Over the course of six weeks, the crew would explore within a range of 250 miles. Their goals were to identify minerals and raw materials that might allow for fuel production and permanent occupation. They would collect Moon rocks, explore craters, and investigate the origin of the Moon. When the mission was complete, they would blast off in their ships toward Earth.

The final issue presented a first mission to Mars, which von Braun said might not happen for a hundred years, given the need for countless new technologies to overcome challenges like duration and cosmic ray bombardment. Still, he offered his best theories in terms of current knowledge, once again with the scale of a twentieth-century invasion. A fleet of ten ships would be assembled in Earth orbit and depart carrying an expedition of seventy scientists and crew members. Traveling at speeds of 80,000 mph, the journey would take eight months each way, though potentially faster with future propellants. Mentioned was a chance for crew hibernation during the voyage.

In Martian orbit, fleet technicians would attach wings to three torpedo-shaped landing planes. The first was equipped with landing skis to touch down on the planet's snowcapped north pole. A 4,000-mile overland journey

would follow, using tractors with inflatable living quarters. Upon reaching the Red Planet's equator, the expedition would construct a base with a landing strip for the other two rocket planes to use. The expedition would spend fifteen months on the surface, studying the planet and awaiting the proper alignment with Earth. With wings removed, the rocket planes would be stood upright to launch off the surface toward home.

The Space Craze

Back on Earth, the response to articles in the *Man Will Conquer Space Soon!* series was amazement. *Collier's* magazine sold millions of copies each week, fueling the imaginations of readers around the country and the world. On the subject of space exploration, science fiction magazines, books, and movies found an ever-growing audience. In the coming years, the *Collier's* articles were expanded into books and television programs.

One fan of the *Collier's* articles was Walt Disney, who produced three related episodes for his weekly show—"Man in Space," "Man and the Moon," and "Mars and Beyond"—which included appearances by von Braun and other authorities. When Disneyland opened in Anaheim in 1955, the Tomorrowland district included a cinematic attraction with moving seats called Rocket to the Moon. Outside the theater stood a 76-foot model of von Braun's Moonliner rocket, sponsored by TWA.

So powerful were von Braun's predictions, many of them remarkably accurate, that they would not only guide American space policy for decades, but also have a profound influence on science fiction. Over a decade later, acclaimed director Stanley Kubrick famously adapted Arthur C. Clarke's *2001: A Space Odyssey*. Widely praised for its fairly accurate depiction of space travel, many of the 1968 film's models bore a striking resemblance to images and ideas from the *Collier's* articles. By the end of the 1950s, von Braun was a household name across America—the German rocket scientist who would soon be asked to win a space race for the Americans against the Soviets.

Meanwhile, in the U.S.S.R., Soviet rocket engineers were following public events in America while working secretly on their own programs. As the head

---

13. A NASA poster created for a 2009 exhibit at the Kennedy Space Center Visitor Complex. NASA/KSC.

of the special design bureau OKB-1, outside Moscow, Korolev's main task was developing a long-range ballistic missile.

The first step was creating a nearly perfect copy of the V-2, which Korolev called the R-1. Next came larger versions, including the 60-foot R-2, which was comparable to the Redstone. This was followed by the 108-foot R-3, which was comparable to the U.S. Air Force's first ballistic missile, the Thor. But Soviet nukes were heavier than American warheads, plus they had to be launched from farther away. So Korolev forged ahead with a more ambitious design for an even larger rocket, one capable of traveling 5,000 miles through space and striking the United States.

6 The Satellites

The International Geophysical Year

Throughout the mid-twentieth century, Cold War tensions deepened. In late 1949 the Chinese civil war came to an end with Mao Zedong establishing the Communist People's Republic of China. Then, in June 1950, the People's Army of North Korea, supported by China and the Soviet Union, crossed the thirty-eighth parallel and invaded South Korea, a Western ally. An American-led UN force intervened, and three years of back-and-forth war across the thirty-eighth parallel led to a draw and restoration of the post–World War II boundaries.

In January 1953 Dwight D. Eisenhower was inaugurated as U.S. president. He pursued a strategy of reducing burgeoning military expenditures by focusing on a more economical policy of nuclear deterrence, mostly relying on long-range bomber aircraft. Less than two months later, Joseph Stalin died. After an internal struggle, Nikita Khrushchev emerged as the new Soviet leader. With a flair for the dramatic, Khrushchev surprised observers at home and abroad. He denounced Stalin's crimes against fellow Soviets and advocated for a more equitable society. Simultaneously, Khrushchev threatened the West.

"Whether you like it or not, history is on our side," Khrushchev told Western diplomats in Moscow. "We will bury you." While Khrushchev later claimed he was speaking symbolically about capitalism, many Americans interpreted this as a nuclear threat.

During this time, communication and collaboration between scientists of the West and East became increasingly challenging. So the International Council of Scientific Unions began planning a worldwide science project to bring politically separated experts together. The International Geophysical Year was planned for July 1957 through December 1958 with the goal of sharing discoveries about Planet Earth. As the project developed, and interest in space exploration grew, a secondary challenge was issued: the launching of satellites to gather scientific data. Both the United States and the Soviet Union agreed to this objective, though neither side moved with much urgency at first.

In the United States, a competitive selection process ensued among the three branches of the military. The air force proposed using the unbuilt Atlas rocket, but their plan was dismissed as hypothetical. Initially, a proposal by the U.S. Army and Wernher von Braun was selected. Called Project Orbiter, it would use the proven Redstone rocket to launch a lightweight satellite. But political blowback soon mounted, with the chief complaints being that the proposed satellite was too simplistic and that the effort should be American only—not led by German scientists. The navy's Project Vanguard, a three-stage vehicle based on the Viking and Aerobee sounding rockets, was given the go-ahead. The goal was to lift a more complex satellite into orbit by spring of 1958.

Sputnik: A Simple Sphere

In the Soviet Union, Korolev initially focused on developing Object D, a 3,000-pound spacecraft that would carry a half dozen scientific instruments. But the complex project soon fell behind. Worried the Americans would launch first, the intensely competitive Korolev pivoted to a much easier design called Simple Satellite 1.

Publicly, the Soviet Union had declared 1958 as their intended launch date. But on the morning of October 4, 1957, a massive R-7 rocket stood on a launch pad in the remote plains of Kazakhstan. A towering 112 feet tall, the triangular rocket was composed of a second-stage central core surrounded by four strap-on boosters comprising the first stage. Together, these two stages of Korolev's engineering marvel had 5 engines with 20 total nozzles. Burning refined kerosene and liquid oxygen, the rocket produced an unprecedented 200,000 pounds of thrust at liftoff.

A few minutes after midnight, it launched with a fiery blaze into the dark sky. A tracking station on the Kamchatka Peninsula picked up the signal. But Korolev waited ninety minutes, until the signal passed over the launch facility, before calling Khrushchev to inform him. A Soviet satellite was in orbit.

In Washington DC, the Soviet embassy was hosting an event related to the

14. At the triumphant post-launch press conference, Pickering, Van Allen, and von Braun display a full-scale replica of Explorer 1. NASA.

International Geophysical Year. A reporter informed the American delegation that the Soviets had beaten the U.S. satellite program into orbit. After the senior American delegate congratulated the Soviets, the entire party went to the roof to try to spot the world's first artificial satellite. It's possible they saw sunlight reflecting off the R-7's upper stage, but it's unlikely they could see the small Sputnik.

The steel sphere, about the size of a beach ball, weighed 184 pounds. It had four long antennae, and the only instrument was a radio, which broadcast pulses every half second. Basically, it beeped. And when Americans woke up the next day and learned it had orbited overhead four times while they slept, it was a shock. If ever the Reds wanted to bury the Americans with nuclear fire from above, they clearly had the technology to do so.

"Soviet Fires Earth Satellite into Space," blared the *New York Times* headline, adding, "Device is 8 Times Heavier Than One Planned by U.S." The Americans talk about spaceflight, one European scientist told a magazine, but the Russians do it. "The new socialist society makes the most daring dreams of mankind a reality," proclaimed the Soviet news agency TASS. The press release indicated more launches were imminent, and they would be larger and heavier.

In Washington DC, President Eisenhower seemed caught off guard—by the launch and the public reaction. Prior to Sputnik, the CIA had considered Soviet claims of launch capabilities to be propaganda. While overflights by the new U-2 spy plane had revealed photos of the R-7 under development in Kazakhstan, the Soviets didn't appear to be ramping up any massive missile systems. Meanwhile, U.S. missile development was perhaps behind but going well.

Five days later, Eisenhower didn't help matters when he downplayed Sputnik as a "small ball." The president mentioned the next test launch of Project Vanguard would happen in December. Eisenhower may have been partially relieved the Soviets had launched first into orbit. A big debate at the time was whether spacecraft had the legal right to orbit over another country's territory. Sputnik had done so without complaints, opening the skies for future satellites.

From Laika to Flopnik

In the Soviet Union, Khrushchev recognized the surprising and immense political capital of Sputnik. He ordered Korolev to launch another satellite within a month to celebrate the fortieth anniversary of the Soviet Revolution. On November 3 another R-7 did just that, and this satellite was even bigger. By hastily combining previous designs, Korolev's second effort was a conical spacecraft weighing 1,100 pounds. A spherical unit like the first Sputnik had been combined with a small instruments package and a barrel-shaped capsule previously tested on suborbital flights.

Inside the compartment, there was a basic life support system and a dog named Laika. Oxygen, water, and food were supposed to last ten days. But due to overheating, Laika probably died sometime during the first few days, which triggered a flood of complaints from around the world. Reflected sunlight made Sputnik 2 periodically visible to the naked eye for months. And any doubts around the world about the Soviets' early superiority in space had been quickly crushed.

In the United States increased pressure fell on the navy to have a successful Vanguard launch. Though it was only a test flight, and the first with all stages active, it was seen as America's answer to the Soviets. After two days of weather postponements, a live TV audience of millions tuned in on December 6, shortly before noon.

The engines fired. Smoke billowed upward from the launch pad. The rocket started to rise. After reaching an altitude of about four feet, it lost thrust and collapsed back onto the pad in a massive explosion. The nose cone came free, and the Vanguard satellite tumbled out of the flames. It was the size of a grapefruit, according to Khrushchev.

"Oh, What a Flopnik," declared the *Daily Herald*, while other news outlets piled on.

"Oopsnik."

"Kaputnik."

"Stayputnik."

"Project Rearguard," wrote *Time* magazine.

Thoroughly embarrassed, and desperate for a success, the United States turned to the army and von Braun. After their satellite proposal had been passed over, von Braun's team had transitioned to modifying the Redstone into the Jupiter intermediate range ballistic missile. In August of 1957, a few months before Sputnik, they had successfully returned the first object from space—a Jupiter nosecone. Fortunately, their team had never actually stopped working on the satellite project. After Sputnik 2, they were authorized to attempt a secret launch. The chance came just before midnight on January 31, 1958.

The rocket that successfully blasted into space was renamed Juno to distance it from military efforts—but it was basically a modified Jupiter C, a version of the Redstone. It was a squat vehicle with four stages, the uppermost being the satellite itself, with a small rocket engine embedded in its missile-shaped casing. Explorer 1 was only six and a half feet long and weighed thirty pounds. Flying into a higher elliptical orbit than expected, it reached 1,575 miles above the earth at its highest point, or apogee.

In a small payload bay, Explorer 1 carried several scientific instruments, including a cosmic ray detector. When mission scientist James Van Allen examined the data, he discovered areas of high radiation that would be confirmed by later Explorer missions. The first U.S. satellite may have arrived late to the party, but it made a major discovery: the Van Allen Belts. During the triumphant press conference, von Braun, Van Allen, and satellite designer William Pickering held a model of Explorer 1 above their heads.

The long-awaited space age had finally arrived. Over the coming months and years, it would manifest as yet another proxy competition between Cold War superpowers, a race in which the Soviet Union remained firmly in the lead.

Sites in New York City and the Northeast
Six Adventures and Ten Orbital Neighbors

Celestial overview: It's slightly less expensive to visit New York City than to travel into space, though it's possible those numbers could switch in the future. In the meantime, New York City offers a pair of excellent museums that should be on any space enthusiast's radar: The Intrepid Sea, Air and Space Museum is more of a classic big artifact facility with an aircraft carrier, the space shuttle *Enterprise*, and more. The American Museum of Natural History is a massive science facility with a planetarium and space-related exhibits about the universe, planets, meteorites, and more. Nearby, on Long Island, the Cradle of Aviation Museum has a solid space gallery.

Elsewhere across the Northeast, the McAuliffe-Shepherd Discovery Center in New Hampshire is a mixed space and science museum. Opening in 2022, the Moonshot Museum in Pittsburgh is an interactive gallery focused on America's return to the Moon, while the nearby Carnegie Science Center developed an immersive exhibition about humanity's future on Mars. Rounding out the region are a series of worthy attractions with mostly smaller space offerings. Some are big science museums with a dash of space, while others are observatories or aviation museums. Each is a bit far-flung from the others, though regional trips may appeal to some enthusiasts.

While New York City and the northeastern United States may not necessarily be considered a top space destination, there's plenty to see for enthusiasts planning a broader visit to the area. When considering the four-hour drive between them, the Washington DC attractions could be combined with the New York City museums into one trip. Or the Pittsburgh museums could be combined with sites in neighboring Ohio. Put it all together, and this region is definitely worth a closer look.

Intrepid Sea, Air and Space Museum, New York NY

Debriefing: The Intrepid Museum is centered around the aircraft carrier *Intrepid*, which is docked in the Hudson River alongside Pier 86 with the museum buildings and aircraft. In addition to tours of duty during World War II and the Vietnam War, the *Intrepid* also served as a recovery vessel for the Mercury and Gemini programs. The space highlight here is the space shuttle *Enterprise*, an atmospheric test orbiter that went on display in 2012. Aviation artifacts include a British Airways Concorde supersonic commercial aircraft, a Lockheed A-12 reconnaissance jet (a precursor of the SR-71 Blackbird), and the *Growler*, a Cold War–era diesel-electric submarine equipped with nuclear cruise missiles.

The STARS: The Space Shuttle Pavilion is centered around the *Enterprise* atmospheric test prototype, plus seventeen exhibits related to the orbiter and the broader shuttle program. Exhibits explore shuttle design, the journey of *Enterprise* from Udvar-Hazy in Washington to New York City, shuttle artifacts, photos, audio recordings, and films. Other highlights include a replica of the Aurora 7 capsule from Project Mercury and a space-flown Soyuz 6 descent capsule.

15. The atmospheric test orbiter *Enterprise* at the Intrepid Sea, Air and Space Museum in New York City. NASA/Bill Ingalls.

Mission parameters: The museum offers regular events like astronomy observations and expert presentations. Check the website for an events calendar.

Countdown: open daily 10:00 a.m.–5:00 p.m.

Mission budget: $$$

Flight team: 877-957-7447 | www.intrepidmuseum.org

Coordinates: 40.7644, -73.999 | Pier 86, West 46th Street, New York NY, 10036

Flight plan: There is no parking at the museum. Visit the museum website for information about private parking facilities and public transportation.

Rations: The Aviator Grill is the museum's on-site family-friendly restaurant.

American Museum of Natural History, New York NY

Debriefing: Located across from Central Park, the American Museum of Natural History includes twenty-six interconnected buildings featuring forty-five permanent exhibitions plus a rotating series of special exhibitions. Exhibition topics range widely, including dinosaurs, forests, ocean life, birds, reptiles, mammals, and the Hall of Human Origins, with archaeological artifacts from around the world.

Space enthusiasts may be particularly interested in two exhibition halls. The Earth and Planetary Sciences Halls include meteorites, minerals, and gems that shed light on the origins of the solar system and properties of Planet Earth. The Rose Center for Earth and Space includes exhibits about the history and properties of the universe. The Rose Center includes the Hayden Sphere, with a lower-level movie theater and upper-level planetarium.

The STARS: Most highlights are in the Rose Center for Earth and Space. The Heilbrunn Cosmic Pathway is a 360-foot spiraling walkway presenting the thirteen-billion-year history of the universe. Another nearby walkway is the 400-foot Scales of the Universe, which illustrates the relative sizes of objects in the universe, from subatomic particles to entire galaxies, with the central Hayden Sphere used to make cosmological comparisons.

Inside the Hayden Sphere is the 429-seat Hayden Planetarium space theater, which offers space shows and a twenty-five-minute film, *Worlds Beyond Earth* (tickets required), about a simulated expedition through the solar system. The lower half of the sphere houses the Hayden Big Bang Theater, which shows a four-minute visualization of the Big Bang.

The Cullman Hall of the Universe features exhibits about the Moon, planets, stars, black holes, galaxies, and the universe. The Gottesman Hall of Planet Earth presents a variety of geologic specimens from across our planet. Located in the Earth and Planetary Sciences Hall, the Ross Hall of Meteorites displays over 130 meteorites and includes exhibits about meteorite impacts, planet formation, and solar system origins.

Countdown: Wednesday–Sunday 10:00 a.m.–5:30 p.m.

Mission budget: $$

Flight team: 212-769-5100 | https://www.amnh.org

Coordinates: 40.7809, -73.9729 | 200 Central Park West, New York NY 10024

Flight plan: There are two museum entrances: on Central Park West at 79th Street and on 81st Street. The museum operates its own paid parking facility, with the entrance on 81st Street between Central Park West and Columbus Avenue.

Rations: There's no restaurant (or food allowed) at the museum, but this is New York City. You'll find something.

Mission parameters: Both the **Intrepid Museum** and **Museum of Natural History** are part of the CityPASS, which includes access to five selected attractions. Since the museums are about 2.5 miles apart, adventurous visitors could walk between them through the southern half of Central Park.

16. Space Park, built for the 1964 World's Fair, is long gone, but the two tallest rockets have been relocated. NASA.

Orbital neighbors: A number of attractions around the city and state of New York have some limited space offerings. Outside the **New York Hall of Science,** at Flushing Meadows Corona Park in Queens, there's an almost-forgotten rocket park. Rising above the landscaped foliage is a pair of shiny launch vehicle mock-ups: an Atlas rocket topped by a Mercury capsule and a Titan II topped by a Gemini capsule. It was originally part of a larger outdoor space exhibit from the 1964 World's Fair. Most of the exhibit fell into disrepair, after decades of neglect, before the few remaining items were restored and moved to their new location. **Flight team:** https://nysci.org.

The **U.S. Golf Association Museum,** about 40 miles east of New York City in Liberty Corner, New Jersey, has the golf club used by Alan Shepard to hit two golf balls across the lunar surface during *Apollo 14.* **Flight team:** www.usga.org/history/visit-the-museum.html.

Cradle of Aviation Museum, Garden City NY

Debriefing: Located on Long Island, this facility has over seventy-five aircraft and spacecraft across eight galleries chronicling the past hundred-plus years of aerospace history. Aviation galleries include a number of nineteenth-century kites and gliders that led the way to developing airplanes. Other artifacts come from the early days of aviation, including a pair of aircraft related to famed aviator Charles Lindberg—a restored Curtiss Jenny he once owned and a replica of the transatlantic *Spirit of St. Louis*. There's also an extensive display of fighters from World War II and the later jet engine era. Space offerings include authentic artifacts, replicas, models, a planetarium, and more.

The STARS: Outside the museum, there's a statue of Sally Ride, the first U.S. woman in space. Inside, a pair of authentic Apollo spacecraft leads the highlights. One is the Command Module 002, which flew in the atmosphere aboard a Little Joe II rocket during a test of the launch-escape system. The other is an authentic flightworthy lunar module from the canceled *Apollo 19* mission. Nearby is the lunar module mission simulator used at Kennedy Space Center, between 1968 and 1972, for Apollo astronauts to practice the procedures for landing on the Moon.

Elsewhere in the gallery, there's the flown spacecraft hatch from *Gemini 2*. One exhibit covers the Hubble and James Webb space telescopes. A number of rockets and missiles includes a full-scale replica of a Goddard A-series rocket and launch stand from the 1930s. There's a Republic JB-2 "Buzz Bomb," an exact duplicate of the Nazi V-1, the first guided missile built for the U.S. Army; a Petrel air-launched anti-ship missile and torpedo; a Republic Terrapin sounding rocket from the 1950s; and a Rigel, an early submarine-launched ramjet missile.

Mission parameters: A planetarium and sky theater offers star shows and films, throughout the day, on a range of topics, including space. Visit the website for showtimes and tickets.

Countdown: Tuesday–Sunday 10:00 a.m.–5:00 p.m.

Mission budget: $$

Flight team: 516-572-4111 | www.cradleofaviation.org

Coordinates: 40.7284, -73.5975 | Charles Lindbergh Boulevard, Garden City NY, 11530

Flight plan: The museum has free parking on-site.

Rations: Restaurants are available throughout the area.

Orbital neighbors:

Located near Schenectady NY, the **Empire State Aerosciences Museum** has some space-related exhibits, including a rocket display and a full-scale replica of the *Apollo* lunar rover. **Flight team:** www.esam.org.

Located in Niagara Falls NY, the **Niagara Aerospace Museum** has a small aviation exhibition and several space artifacts, including an Apollo lunar module ascent-stage test engine. **Flight team:** https://niagaraaerospacemuseum .org.

Located in Charlestown RI, the **Frosty Drew Observatory & Science Center** offers free telescope viewing to the public. **Flight team:** https:// frostydrew.org/observatory/.

Located in Tupper Lake NY, the **Adirondack Sky Center** is a telescope observatory offering nighttime sky viewing with expansion plans for a science museum and planetarium. Check the website for updates. **Flight team:** www.adirondackskycenter.org

McAuliffe-Shepard Discovery Center, Concord NH

Debriefing: The McAuliffe-Shepard Discovery Center is named for two pioneers in U.S. space exploration originally from New Hampshire. Christa McAuliffe was selected to be NASA's first teacher in space but tragically died during the 1986 space shuttle *Challenger* explosion. Alan Shepard was the first American in space, during Project Mercury, and he walked on the Moon during the *Apollo 14* mission. The center mixes tributes to its honored namesakes with a variety of space- and science-related exhibits.

The STARS: Standing in a plaza outside the museum, there's a full-size replica of the Mercury-Redstone rocket. Surrounding panels tell the story of New Hampshire astronaut Alan Shepard, who became the first American in space in 1961.

Inside the museum, there's a full-size replica of the Mercury capsule that took Shepard into space. The exhibit To the Moon! focuses on lunar exploration, science, and future missions. Another exhibition focuses on the Lunar Reconnaissance Orbiter and its discoveries about the Moon.

Space shuttle exhibits include a large-scale model and artifacts like a space toilet, test mission ejector seat, prelaunch bolt, space treadmill, and shuttle tire. Other exhibits focus on the Sun, human physiology in space, the planets, and the solar system.

EVAS: The center's planetarium offers several showtimes throughout the day, with tickets available online.

Countdown: April 15–May 1 and June 17–September 4: open daily 10:30 a.m.–4 p.m.; March 7–April 14 and May 2–June 16: open Friday–Sunday 10:30 a.m.–4:00 p.m.; closed most holidays.

Mission budget: $

Flight team: 603-271-7827 | www.starhop.com/

Coordinates: 43.2241, -71.533 | 2 Institute Drive, Concord NH 03301

Flight plan: The center is located on the north side of Concord, near the junction of I-93 and I-393.

Rations: The center's Countdown Café is located on-site, with more restaurants available in town.

Orbital neighbors:

About an hour north of the center, in the tiny town of Warren, an authentic

Redstone missile stands in front of the **Warren Historical Society Museum**. Flight team: 603-764-5869.

Located in Boston, the **Museum of Science** has several space-related exhibits, including exhibits about the Moon, solar system, and galaxy. There's also a planetarium and theater. **Flight team:** www.mos.org.

Located in Ottawa, the **Canada Aviation and Space Museum** is a large aerospace museum with an impressive exhibition of over 130 aircraft and several space-related exhibits. Artifacts include the space-flown Canadarm from the *Endeavor* space shuttle, plus other displays presenting the country's space achievements, including space health discoveries. A Life in Orbit exhibit about the International Space Station includes replica modules like the multiwindow Cupola. **Flight team:** https://ingeniumcanada.org/aviation.

Moonshot Museum, Pittsburgh PA

Debriefing: Opening in 2022, this unique experience-based museum focuses on current and future space missions with an emphasis on NASA's return to the Moon. Interactive exhibits and presentations range from simulated lunar missions to career preparation for the contemporary space industry. Located in the same building as aerospace company Astrobotic Technology, the museum offers insights into the company's ongoing lunar efforts. Other topics include microgravity research and future exploration of the solar system.

The STARS: The museum is divided into experience zones, beginning with an entryway theater showing an introductory film about the contemporary space industry.

A wall of windows allows views into the Astrobotic clean room workshop. Visitors may see efforts related to the company's Artemis projects, including the Griffin lander, which will deposit the *Viper* rover on the lunar surface to search for frozen water.

Elsewhere, simulated environments will allow visitors to consider the topography and geology of the lunar surface, including lava tubes where humans may search for future settlement sites. An interactive lunar lab blends tactile and digital elements, allowing visitors to assemble and test a prototype lunar rover.

A simulated habitat will envision a lunar settlement many decades in the future, with activities including discussions of the sociocultural considerations of living and working on the Moon. Other interactive topics may include establishing entrepreneurial space companies or solving a mystery on the Moon.

Displays will include a scale model of the Vulcan Centaur rocket, which is slated to launch Astrobotic's Peregrine lander to the Moon in 2022. A variety of public tours and educational programs will become available as the museum develops.

Countdown: TBA

Mission budget: $

Flight team: 412-314-4111 | https://moonshotmuseum.org/

Coordinates: 40.4503, -80.0201 | 1016 North Lincoln Avenue, Pittsburgh PA, 15233

Mission parameters: At the time of research, the museum was finalizing precise programming, hours, and parking options, subject to change. Contact the museum or visit their website for updates.

Rations: Restaurants and stores can be found throughout the city.

Carnegie Science Center, Pittsburgh PA

Debriefing: One of four Carnegie museums in Pittsburgh, the Carnegie Science Center is an impressive facility with interactive exhibits and displays on local rivers, miniature railroads, robots, energy, and more. Moored outside, in the Ohio River, the USS *Requin* is a World War II–era submarine that visitors can board. The center has two space-related exhibits, a replica International Space Station module, and an immersive exhibition about future missions to Mars. There's also a planetarium, theater, and telescope observations.

The STARS: During 2022 the center was finalizing designs for a new exhibition, Mars: The Next Giant Leap. Organized into seven zones, this interactive exhibit will take visitors on a journey to a hypothetical human future on Mars. Large images and wall screens will simulate various phases of the operation, like the voyage to the Red Planet, plus observations made from Martian orbit and views of Earth from Mars. Displays will illuminate the efforts of Pittsburgh aerospace companies to prepare for space missions, eventually leading to a voyage to Mars, including contemporary artifacts like CubeSats, a common type of miniature satellite, and other observational objects. Human perceptions of Mars throughout the course of history will be presented.

Other zones will consider issues like Martian climate and how observations made on the Red Planet may reshape humanity's understanding of Earth. An interactive display will allow visitors to manipulate Martian climate variables and see the planet-wide results. Other displays will focus on the Martian landscape, including the role of rovers in searching for water. Visitors will be able to remotely control miniaturized Mars rovers in a simulated landscape. Artifacts will include an authentic Martian meteorite and Moon rock.

A garden area will envision an enclosed greenhouse habitat, where fruits and vegetables are grown. Another zone will simulate the interior of a Martian homestead, where household objects, audio elements, and fictionalized settlers can be discovered. A final gallery will simulate a Martian cave adapted for human use and feature a miniaturized model Martian settlement that will expand over time with input from guests during activities.

Elsewhere in the museum, an exhibit about the International Space Station includes a module mock-up, astronaut artifacts, and displays.

EVAS: The Buhl Planetarium offers shows and films, included with general admission, about the stars and solar system. A telescope observatory offers nighttime viewing for an extra fee. Check the website for current schedules.

Countdown: Wednesday–Monday 10:00 a.m.–5:00 p.m.; closed Tuesday

Mission budget: $$

Flight team: 412-237-3400 | https://carnegiesciencecenter.org

Coordinates: 40.446, -80.018 | One Allegheny Avenue, Pittsburgh PA 15212

Flight plan: The center is on Pittsburgh's North Shore of the Ohio River, not far from downtown. Paid parking is available on-site.

Rations: The **RiverView Cafe** is the center's on-site restaurant.

Orbital neighbor: Located in Philadelphia, the **Franklin Institute** is a science and education center named after early American scientist and statesman Benjamin Franklin. The center includes exhibits about electricity, machines, trains, engineering, the human brain, and Newton's laws of motion. The Space Command exhibition includes an *Apollo 15* Moon rock, a meteorite to touch, and several spacesuits. Displays cover the solar system, space shuttle program, International Space Station, and more. There is a gallery of antique telescopes, and a planetarium and domed theater offers star shows and films. **Flight team:** www.fi.edu.

7 The Space Race

Object D

In mid-March of 1958, Project Vanguard finally succeeded in launching the second American satellite, after a second failure the previous month. This truly small ball, a quarter the size and one-sixtieth the weight of Sputnik, did achieve a milestone—the first spacecraft to use solar panels. That year, the United States managed to get two more Explorers into orbit against the failures of two other Explorer launches and four more Vanguard failures.

Once again, U.S. efforts were overshadowed. In mid-May, the Soviets finally launched Korolev's Object D, which was officially designated as Sputnik 3. Though appearances suggested otherwise, the Soviets were also experiencing their share of failures. The previous month, the original Object D had been destroyed during a launch failure. But Soviet secrecy meant they could conceal most miscues. Since the state controlled the press, they only publicized successes. A backup was fit atop the increasingly reliable R-7.

Sputnik 3 was massive. At just under 3,000 pounds, it was a hundred times heavier than the U.S. Explorers. The conical Soviet vehicle carried twelve scientific instruments, powered by batteries and solar panels. It was a true spacecraft, essentially a laboratory in space. Twelve feet long by six feet wide at its base, you could fill up the instruments bay with American satellites and have plenty of room to spare. The Soviets put a replica on display in Moscow for the world to see.

From NACA to NASA

Back in the United States, it was becoming clear that the piecemeal approach, with different programs spread out among competing agencies, wasn't working. After internal discussions, Eisenhower decided a major reorganization of all U.S. space efforts was needed. The military branches would retain all

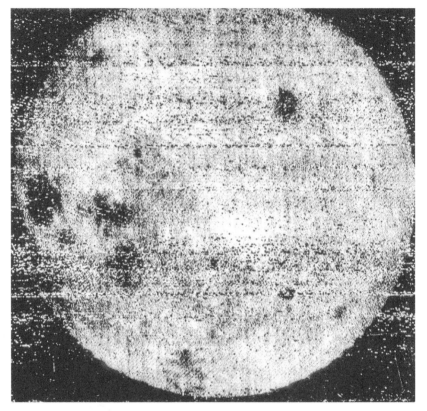

17. Humans' first glimpse of the far side of the Moon was a grainy photo taken by the Soviet probe *Luna 3* in 1959. OKB-1.

space weapons projects, such as ballistic missiles. The air force would focus on long-range vehicles and the army and navy on various shorter-range projects. Meanwhile, a single agency was needed to oversee all civilian space projects.

The logical answer was the National Advisory Committee for Aeronautics. Founded in 1915 to promote aeronautical research, NACA had developed wind tunnel research and key technologies for military planes during World War II. After the war, it had helped develop the Bell X-1, a rocket-powered airplane shaped like a bullet with straight wings. In 1947 Chuck Yeager piloted the X-1 past the speed of sound, the first manned aircraft to do so in level flight. More recently, NACA had commissioned a design from former V-2 leader

Walter Dornberger for a hypersonic research aircraft: the rocket-powered X-15, which would fly to the edge of space.

In July 1958 NACA was expanded into the new National Aeronautics and Space Administration. Within a year, NASA had assumed control of every civilian space project in the U.S. government. The Army Ballistic Missile Agency at Redstone Arsenal became NASA's General George C. Marshall Space Flight Center, with Wernher von Braun named director.

Given the Soviets had won the race for a satellite, NASA's early tasks would focus on the next major milestones: launching an American into orbit and sending unmanned spacecraft out of Earth's orbit.

The first priority was Project Mercury, involving the selection and training of space pilots. After a spacecraft was designed and built, it would be launched on suborbital flights using the reliable Redstone rocket. Next, the air force's more powerful Atlas ICBM, which was nearing completion, would be adapted to launch a Mercury capsule into orbit. The plan seemed sound, and this time the Americans might have a chance at being first. Since Sputnik 3, the Soviet space program had seemingly gone quiet. Though they were certainly hard at work behind the scenes.

First to the Moon

The first attempts to send a probe to the Moon were American. In August 1958 the U.S. Air Force launched a Thor-Able three-stage rocket that rose for seventy-seven seconds before exploding. Later that year, NASA took over the program and tried again in October. This time the three-stage rocket reached an apogee of over 70,000 miles high—about one-quarter of the way to the Moon. But the angle was too steep, and the probe plummeted back to Earth, landing in the Pacific Ocean after forty-three hours. Officially recognized as *Pioneer 1*, it was called a magnificent failure for briefly setting a new altitude record.

The same hardware was launched again in November, but the solid-fueled third stage failed, and *Pioneer 2* reached only 963 miles altitude during a forty-one-minute flight. For *Pioneer 3*, in December, they tried a Juno II—one of von Braun's variations on Jupiter and Redstone with upper stages built by NASA's Jet Propulsion Laboratory. This one made it 63,000 miles up before plummeting after thirty-three hours.

Meanwhile, the Soviets were trying to reach the Moon as well. During the fall of 1958, they made three attempts using an upgraded R-7 rocket with a new upper stage, but each exploded. In early January of 1959, they succeeded.

Luna 1 became the first spacecraft to reach escape velocity of just over 25,000 mph and depart Earth's orbit. Weighing around 800 pounds, the craft was a steel sphere about three feet in diameter with a half dozen antennae. The true objective, which the Soviets didn't reveal, was to crash into the Moon's surface. But due to a ground control error, the third stage burned too long. *Luna 1* missed the Moon by about 3,700 miles and became the first human-made object to enter a heliocentric orbit, circling the Sun between the orbits of Earth and Mars. The Soviets had yet another triumph. The probe was given the nickname Mechta, meaning Dream, and the government released commemorative stamps to mark the event.

Two months later, the United States responded. In early March 1959, a Juno II rocket achieved a perfect launch from Cape Canaveral, escaping the earth's gravity and sending the first American spacecraft toward the Moon. *Pioneer 4* was a conical probe weighing only thirteen pounds. And it missed the Moon by 36,000 miles, ten times the distance of *Luna 1*. The probe carried a camera, but it never got close enough to take a photo. The press, the public, and the Soviets shrugged. No one cared.

Like they had with the Sputniks, the Soviets soon piled on. In midsummer, they concealed a failure by purposefully exploding an errant rocket during launch. But in September they struck again with another lunar probe: *Luna 2*, which successfully crashed into the Moon. Onboard was curious cargo designed to explode on impact. Two titanium spheres resembling soccer balls, each composed of seventy-two pentagonal panels, which the Soviets called pennants. Each pennant was engraved with the name and emblem of the U.S.S.R. When *Luna 2* reached its target, these small Soviet victory flags were designed to scatter across the lunar surface.

A Soviet Pennant

The very next day, none other than Khrushchev arrived in the United States for the first-ever state visit by a Soviet leader. As a gift for President Eisenhower, Khrushchev brought an identical titanium pennant sphere.

18. The far side of the Moon as photographed during the *Apollo 16* mission in 1972. NASA.

"A pennant bearing the national emblem of Soviet Union is now on the Moon," said Khrushchev during his arrival remarks. "We entertain no doubt that the splendid scientists, engineers, and workers of the United States of America, who are engaged in the field of conquering the cosmos, will also carry their pennant to the Moon. The Soviet pennant, as an old resident of the Moon, will welcome your pennant and they will live together in peace and friendship."

About a week after Khrushchev went home, the Soviets successfully launched *Luna 3*. This egg-shaped spacecraft used solar panels to power a sophisticated camera that took the first photos of the Moon's far side. Because the Moon is tidally locked, the same familiar side always faces Earth. Thus, this was the first time humans had ever glimpsed the hidden face of the Moon. The images were grainy, but they clearly showed the far side was very different from the near side. The far side was seemingly mountainous and mostly lacking the vast lunar seas called marias. First with satellites and now with Moon probes, the Space Race was firmly in the Soviets' hands.

8 Mercury vs. Vostok

The First Astronauts

Named after the Greek messenger god, Project Mercury was NASA's first big assignment. Place an American in space as soon as possible. Development of a small one-person capsule began immediately. But who would get to ride inside?

One suggestion was an open call for any American to apply, including civilians. Another scientist suggested American women, given their smaller size, need for less oxygen, and ability to cope with stress. Another proposal was military test pilots, all of whom were men during the late 1950s. Eisenhower decided on the latter group with the stipulation they be active duty only. These stipulations immediately disqualified countless interested individuals, including NASA's civilian X-15 pilot Neil Armstrong, who had been selected for the air force's now-canceled program Man in Space Soonest.

NASA drew up a posting, in the process popularizing a relatively new job title: astronaut. Soon, they were winnowing the list down from five-hundred applicants. In a diplomatic move, they selected three candidates from the air force, three from the navy, and one from the marines. NASA introduced their first astronauts at a press conference on April 9, 1959. They were called the Mercury Seven, and when a reporter asked who wanted to go first, each of them raised a hand. Except for John Glenn and Wally Schirra, who raised both hands.

Overnight, they became American heroes. Here were the boys that were going to take on the Soviets for supremacy in space. For PR purposes, and to boost their meager military salaries, NASA arranged for the seven men to sell their stories to *Life* magazine. The publicity was universally positive. The astronauts posed for group photos in their shiny silver spacesuits. They studied engineering, space science, and astronomy. They underwent survival training in the desert. They made parachute jumps and spent hours in

19. The *Mercury 7* astronauts. NASA.

simulators. They flew on the Vomit Comet, a C-131 that followed a parabolic flight plan to simulate weightlessness.

Meanwhile, work continued on the Mercury spacecraft. During the design stage, the reentry capsule—where the single astronaut would sit—evolved from a simple cone to a more bell-shaped profile. Early mock-ups were almost entirely automatic, giving the astronauts little to do but sit there. When the accomplished test pilots complained, flight controls were added for manual override.

Results from the early rocket tests, however, were mixed. The first of three phases involved eight uncrewed test flights with the Little Joe, a 55-foot-tall two-stage solid rocket. The Little Joe launched boilerplate capsules, test articles that collected flight data, topped with launch-escape towers that could pull the astronaut to safety should the booster malfunction. During the first test launch, the abort rocket fired while still on the launch pad. During later flights, the abort system fired too late or too early, and one time it didn't separate the capsule from the rocket as intended. But other escape tests did succeed, including two flights with rhesus monkeys named Sam and Miss Sam, with both animals recovered safely.

A Four Inch Spaceflight

The second phase of Project Mercury was called Mercury-Redstone, during which von Braun's rocket would be used to launch the capsule on a suborbital flight. Mercury-Redstone 1, the first test launch of the 84-foot liquid-fueled vehicle, was scheduled for late November from Cape Canaveral. Recovery forces waited in the Atlantic, while the press gathered at the launch site.

The engine ignited at 9:00 a.m. The control gantry swung away, and the rocket began to rise. A total of four inches. Then it settled back onto the launch pad. The escape tower rocket jettisoned, leaving the capsule behind. A few seconds later, the parachutes popped out one by one. And the rocket—filled with flammable alcohol and liquid oxygen—just stood there, unsecured to anything. The parachutes hung by cables, threatening to catch wind on what was, luckily, a calm weather day.

The control room was at a loss for how to proceed with defusing this massive and precarious bomb. One technician suggested depressurizing the liquid oxygen tank by shooting it with a rifle. The flight director, Chris Kraft, rejected this idea. Instead, the team waited for hours while the batteries ran down and the oxidizer boiled off.

It was called the Four Inch Flight. A major embarrassment. Project Mercury's own Flopnik. The reason for the misfire was a simple electrical flaw in some ground cables. Fortunately, the program rebounded. The next Mercury-Redstone test flight was completely successful, lofting the empty capsule 130 miles high to be recovered by helicopters 235 miles downrange in the Atlantic Ocean.

The third flight carried Project Mercury's first passenger, a chimpanzee named Ham. The goal was to certify the spacecraft for human use. On the last day in January 1961, the spacecraft launched. The rocket flew faster and higher than planned, subjecting Ham to around 15g—fifteen times the force of gravity. Despite his discomfort, he survived to become the first primate in space. After landing in the Atlantic, about 120 miles from the recovery fleet, Ham was rescued just before his capsule sank. Due to the booster mishap, von Braun insisted another test flight was necessary before risking a human flight. The astronauts and other NASA officials complained, but the test flight was authorized and scheduled for late March.

Closely watching these events was a new figure in the White House, Pres. John F. Kennedy. Throughout the 1960 election, he'd campaigned aggressively against the Eisenhower administration on a wide variety of issues, including several related to space. One was a perceived missile gap with the Soviet Union, which appeared to have a greater ability to deliver nuclear weapons. Another topic was complacency around space travel, including the Soviet victories with the Sputniks and the *Luna* probes. If elected, Kennedy vowed to devote the necessary resources to missile and space development. In November, he had soundly defeated Eisenhower's vice president, Richard Nixon. Now Kennedy hoped to reap the political benefits of putting the first man in space.

The Sputnik that Hit Wisconsin

During an early September morning in 1962, a pair of police officers were on patrol in Manitowoc, Wisconsin. As they approached the Rahr-West Art Museum on North 10th Street, they spotted a round hunk of metal embedded in the middle of the street. It was shaped like a small hubcap, maybe eight inches in diameter. When they circled back to take a closer look, they found the object was too hot to touch.

It would later be identified by NASA as a piece of Sputnik 4. A larger satellite, the Soviets claimed, which spent twenty months in a decaying orbit before breaking up over the shores of Lake Michigan. But the truth was much more remarkable. Though the officers didn't realize it, the hunk of metal they kicked out of the street that morning was a piece of the Soviets' first Vostok spacecraft.

In progress since early 1957, the *Vostok* represented several years of secretive work by Soviet chief designer Sergei Korolev. The spacecraft mated a conical instrument module, based on Sputnik 3, with a spherical reentry capsule. Inside this capsule, there was a single seat to carry the first Soviet cosmonaut into space.

20. Cosmonaut Yuri Gagarin on the bus before launching on *Vostok 1*. NASA.

Once the spacecraft was in production, it was time to find some occupants. Cosmonaut selection teams spread out across the Soviet Union. As with the Americans, the focus was on accomplished fighter jet pilots. From three thousand interviewees, they selected around a hundred candidates. Then they put them through rigorous tests, including days in isolation chambers. When the group was down to twenty, they reported for training in Moscow. Gen. Nikolai Kamanin's regimen involved academic lectures and physical training, including parachute jumps, parabolic plane flights and, curiously, ejector seat training. The group was further whittled down to six, and then two, who were filmed during a staged meeting of the Soviet State Committee for later release. The choice was Yuri Gagarin, the son of peasant farmers from western Russia, who had become an engineer and fighter pilot.

The Flight of Gagarin

In Kazakhstan, Korolev woke Gagarin and his backup at 5:30 a.m. on April 12, 1961. They donned their orange spacesuits and boarded a bus for the nearby Baikonur Cosmodrome. Gagarin was strapped snugly inside the small capsule atop an R-7 rocket. It was designed to fly on autopilot; the simple control panel had only four switches and thirty-five indicators, with a hand controller for emergencies. The only way Gagarin could assume control was with a four-digit code, to be sent from the ground during an emergency. At least four people whispered the code to Gagarin before the hatch was sealed.

The countdown began and, at 9:07 a.m., the R-7 roared into the sky. After two minutes, the four strap-on boosters fell away at right angles to one another—a stunning visual later nicknamed the Korolev Cross. Three minutes later, the rocket core stage disengaged. The payload shroud opened and the final third stage pushed the *Vostok* into orbit. Korolev had done it again.

"I can see the Earth," said Gagarin, looking through the porthole window. "I continue the flight—everything is good."

When the third stage detached, the spacecraft was over northern Siberia, traveling 17,500 mph toward a max altitude of 177 nautical miles above sea level. Beneath passed the Kamchatka Peninsula. Sunset came thirty minutes into the flight, as the vehicle crossed the Pacific Ocean. The Hawaiian Islands appeared below as tiny specks of light. The Sun rose again thirty-three minutes later, just after passing Cape Hope at the tip of South America. In total, the mission took 108 minutes to circle the earth.

The retrorockets, which slowed the craft for descent, fired above the West Coast of southern Africa. But as the descent and instrument modules separated, the main electrical cables remained attached. The tethered modules began to spin wildly. Finally, the heat of reentry burned through the wires, and the descent module righted itself. Gagarin struggled to stay conscious while experiencing forces around 8g. The descending spacecraft passed over the Black Sea and reentered Soviet airspace. At around 22,000 feet, automatic sensors blew the hatch and the cosmonaut's seat ejected—a secret about the flight that the Soviets kept until 1971. Gagarin parachuted toward the ground, while the descent module's parachutes opened as well.

About 900 miles northwest of the launch site, on the ground in Saratov Province, a farmer and her granddaughter watched an orange figure wearing a large helmet parachute into a field. Not far away, a smoking metal sphere slammed the ground and bounced. As Gagarin walked toward them, the witnesses backed away in fear.

"Don't be afraid," yelled Gagarin, after removing his helmet. "I am a Soviet like you, who has descended from space, and I must find a telephone to call Moscow!"

Gagarin was awarded the nation's highest honor, Hero of the Soviet Union. The largest celebrations since the end of World War II were held across the U.S.S.R., and the cosmonaut was paraded through adoring crowds.

21. President John F. Kennedy addresses a joint session of Congress to issue his Moon landing challenge. NASA.

In the Thick of It

Kennedy and Catching Up

In the United States, the headlines about the Soviets' latest space accomplishment surprised no one. The first man in space would forever be Yuri Gagarin, not an American. President Kennedy sent congratulations to Khrushchev for the "outstanding technical achievement." During a press conference, a reporter asked Kennedy when the United States might equal the Soviets in space.

"We are behind," Kennedy admitted. "The news will be worse before it is better, and it will be some time before we catch up."

Behind the scenes, like many Americans, Kennedy was fuming. Two days later, he summoned his top advisers to a meeting in the Oval Office. NASA's deputy administrator told the president that matching the Russians might require a program similar in magnitude to the Manhattan Project, which had produced the atomic bombs. "If someone can just tell me how to catch up," said Kennedy. "There's nothing more important."

Kennedy assigned Vice Pres. Lyndon Johnson the task of identifying "a space program which promises dramatic results in which we could win." One of the experts Johnson approached was Wernher von Braun, who had a lot of say on the matter. Intelligence reports suggested the ambitions of the secretive Soviet program's chief designer went far beyond that which they'd publicly achieved.

In October of 1960, the Reds had launched two probes toward Mars, timed to coincide with Khrushchev's second visit to the United States. Both had failed during launch and were concealed. Then, in February, the Soviets sent two probes toward Venus. The first didn't leave Earth orbit, so they conveniently labeled it Sputnik 7. But a week later, *Venera 1* successfully departed for Venus. The Soviets had publicized this occurrence, but contact was lost after a week, so the mission received little attention. Even now, the probe was closing in on the solar system's second planet, where it would ultimately pass within 60,000 miles.

Von Braun had extrapolated from these endeavors that the Soviet R-7 could place 14,000 pounds of payload in Earth orbit. In contrast, the Project Mercury capsule weighed only 3,900 pounds. Basically, the Soviets could soon launch a multiperson spacecraft. Or they might launch a space station. Or perhaps they might attempt a soft landing on the Moon. Or maybe even send a crew around the Moon. And all these things could be done before the United States.

But there was one exception. Landing a crew on the Moon and returning them safely to Earth would require a rocket ten times bigger than what the Soviets were currently using. The United States had no such rocket, but it was unlikely the Soviets did either. In fact, U.S. spy satellites had revealed a massive explosion at the Baikonur Cosmodrome back in October 1960, one that destroyed a launchpad and killed an unknown number of personnel, suggesting they were struggling more than people realized. Von Braun believed there was only one solution. He claimed with a crash program it could be accomplished by late 1967 or 1968—a cunning timeline that fell within the president's potential second term. Johnson took the information back to Kennedy.

An American in Space

A few weeks after Gagarin's orbital flight, Alan Shepard was strapped into a Mercury capsule atop a Redstone rocket. "Why don't you light the damned candle," blurted Shepard, annoyed after days and now hours of delays. "I'm ready to go!"

With even further delays, Shepard soon encountered another problem. He had to pee. The engineers hadn't planned for this. So the power was shut off and he went to it.

Freedom 7 was just a suborbital flight lasting fifteen minutes. But NASA broadcast the entire event on television, and forty-five million Americans tuned in. Given this success, it seemed like the United States was finally starting to catch up.

Three weeks later, Kennedy made an address to a joint session of Congress. "I believe that this nation should commit itself to achieving the goal, before this decade is out, of landing a man on the Moon and returning him safely to the earth. No single space project in this period will be more impressive to mankind or more important for the long-range exploration of space, and none will be so difficult or expensive to accomplish." Americans hadn't even reached orbit yet, but now they were setting their sights on the Moon. The date was May 25, 1961, and there was a whole lot of work to be done.

Meanwhile, Mercury and Vostok flights continued. In late July, Gus Grissom flew in *Liberty Bell 7* on another suborbital hop. The flight went perfectly, but upon landing, the escape hatch blew prematurely. The capsule began filling with water, and Grissom had to swim away as the spacecraft sank into the Atlantic.

A few weeks later, *Voskhod 2* launched with Gherman Titov aboard. During a full-day flight, Titov orbited the planet seventeen times. This gave the Soviets their first chance to examine the effects of extended weightlessness on the human body. Eating and drinking went fine, but when Titov tried to sleep, he became nauseated.

John Glenn and Katherine Johnson

The following February, in 1962, Project Mercury was ready for its third phase, an orbital flight. John Glenn was strapped into the *Friendship 7* spacecraft atop a shimmering stainless-steel Atlas rocket. While securing the hatch, one of the seventy bolts was broken, so the entire process had to be restarted. Other delays related to the math, specifically calculating the orbital trajectory for the flight.

During the early days of the space age, before electronic computers took over, the space program employed a large workforce of human computers. Many of these critical mathematicians were women, including African Americans, and the teams were racially segregated until 1958, when NASA formed. By the time Project Mercury was ready to launch, segregation may have officially ended but much discrimination continued, and the important contributions of these women remained hidden from sight. One of NASA's top mathematicians was an African American woman named Katherine Johnson, and Glenn refused to launch until she personally checked the math for the mission.

Finally, it was time. Shortly before 10:00 a.m., Glenn roared into the sky. The flight path took him across the Atlantic, southeast past Africa, and

22. Katherine Johnson at her desk in 1966. NASA.

into the night side of Earth over the Indian Ocean. As the Sun rose beyond Australia, Glenn reported something odd outside the window. "I am in a big mass of some very small particles, that are brilliantly lit up like they're luminescent . . . They look like stars. A whole shower of them coming by."

The flight continued, northeast over the Pacific and across Mexico and the southern United States. The mysterious cloud, of what was later described as "fireflies," reappeared several times during Glenn's flight. He reached an altitude of 165 miles at an orbital speed of 17,500 mph. After automatic control thrusters malfunctioned, Glenn took manual control of the spacecraft. When a warning sensor indicated the heat shield had come loose, ground control feared the "fireflies" might be pieces of the spacecraft disintegrating.

After three orbits and just under five hours in space, the decision was made to bring Glenn down with the retrorocket pack still attached. Strapped to the heat shield, it might hold the capsule together for reentry. During the fiery descent, Glenn watched flaming pieces of the straps fly past the window. After a successful splashdown, tests showed the warning sensor had been faulty. Glenn was greeted as an instant hero, and four million people lined the streets of New York for a parade in his honor.

Mercury's next mission was another three-orbit flight. Scott Carpenter photographed the earth, performed a short experiment to see how fluids behaved in weightlessness, and solved the mystery of the "fireflies." Harmless ice crystals dislodging from the spacecraft exterior. One glitch was a guidance error, which caused him to land 250 miles from the recovery crews. By the time they arrived, he'd been floating in a life raft for hours.

Reconnaissance and Crisis

Since summer of 1961, Soviet space efforts had been relatively quiet. Behind the scenes they were devoting their R-7 launches to the development of an uncrewed spy satellite, the Zenit. Based upon the Vostok spacecraft, the Zenit carried cameras pointed at the earth. After several weeks, it landed in Soviet territory.

The Zenit project was an answer to one space endeavor where the United States was leading, orbital reconnaissance. After the Sputnik crisis, Eisenhower had authorized the Corona program, which became operational by mid-1959. When an American U-2 spy plane was shot down over Soviet territory in 1960, the orbital program expanded. The Corona spy satellites were mostly missile-shaped vehicles boosted by Thor-Agena rockets into polar orbits. They each carried a single camera, and the film was ejected in a reentry capsule. After a parachute was deployed, these film canisters were retrieved midair by patrolling air force planes that trailed custom hooks. It was these satellites that obtained the detailed images of Soviet territory that revealed the R-7 and Baikonur catastrophe in 1960.

In mid-August of 1962, the Soviet crewed space program roared back to life

23. Project Mercury-Redstone rocket launches Alan Shepard aboard *Freedom 7* in 1961. NASA.

with two launches a day apart. On August 11, Andriyan Nikolayev launched in *Vostok 3*. A day later, Pavel Popovich joined him in orbit, marking the first time two people had been in space at once. When they returned to Earth on August 15, Nikolayev had set a record—four days in space.

In early October, Project Mercury launched Wally Schirra on a six-orbit mission. He tested automatic control systems, used elastic exercise devices, and made the first live TV broadcast from space.

A few weeks later, a different type of rocket caused an international incident. Reconnaissance photos from a U-2 spy plane showed the deployment of Soviet nuclear missiles in Cuba, only 90 miles from Florida. Khrushchev's move was a response to two U.S. actions. In 1961 the failed Bay of Pigs Invasion, orchestrated by the CIA and executed by Cuban exiles, had failed to overthrow Cuban revolutionary leader Fidel Castro. That same year, the U.S. Air Force began deploying the Jupiter missile, designed by von Braun and armed with nuclear warheads, in Italy and Turkey. In the summer of 1962, Castro had requested the placement of nuclear missiles and Khrushchev had agreed. The Soviets sent R-12s and R-14s, both recent derivatives of Korolev's R-series rockets, designed by Mikhail Yangel.

In response, Kennedy's military advisers recommended an air strike and invasion. The president opted for a more cautious approach: a blockade to prevent further deployments, followed by a demand to remove all Soviet weapons. During tense negotiations, Khrushchev agreed on two conditions. The United States publicly declared it would not invade Cuba, while it secretly removed its Jupiter missiles from Italy and Turkey. It was the closest the two superpowers had come to nuclear conflict.

The following year, during May of 1963, the final Mercury mission launched. It was called *Faith 7*. Gordon Cooper completed twenty-two orbits in thirty-four hours, even launching a small satellite during his flight.

The Soviets followed with one final Vostok group flight in mid-1963. After Valery Bykovsky launched, two days later, Valentina Tereshkova became the first woman to travel into space. Bykovsky set a new record, seventy-seven orbits and five days—a solo record that stands to this day. Tereshkova logged forty-eight orbits over three days. When they landed by parachute, local villagers rushed out to help and invited Tereshkova to dinner. She became a worldwide celebrity and, like Gagarin, was honored as Hero of the Soviet Union. Thus, the Vostok program ended in spectacular fashion.

Over at NASA, some mused about sending up a three-day Mercury mission. But the consensus was to move on to bigger and better things. Since Kennedy had issued his Moon challenge two years before, a tremendous amount of progress had already been made. Next, NASA would proceed to Project Gemini, a two-person spacecraft designed to test spaceflight techniques needed for the concurrently developing Apollo program.

Sites Near Cape Canaveral, Florida

Three Adventures and One Orbital Neighbor

Celestial overview: Marketed as the Florida Space Coast, the region surrounding Cape Canaveral is home to several space age adventures plus space-themed landmarks and sites for viewing rocket launches. By far, the top highlight is the Kennedy Space Center Visitor Complex, a massive museum and amusement park with countless space-flown artifacts, replicas, and exhibits that should not be missed. While in the area, there are two smaller space museums nearby worth checking out, and a large science museum not far away in Orlando.

   Launch viewing-Mission parameters: Some enthusiasts plan their visits to Cape Canaveral around scheduled rocket launches. Important to note is that launches are frequently delayed, often at the last minute, due to weather or technical issues. Launch viewing locations can be found across the Space Coast. One popular option is Kennedy Space Center Visitor Complex, which offers several viewing sites that are among the closest to the pads. The center maintains a web page with a calendar listing upcoming launches and a countdown to the next launch. Sometimes launch viewing is included with daily admission, in which case it's strongly recommended to arrive early for the limited spaces. At other times, advance tickets may be required for launches and can be purchased online. **Flight team:** www.kennedyspacecenter.com /launches-and-events/events-calendar/see-a-rocket-launch.

Kennedy Space Center Visitor Complex, Merritt Island FL

Debriefing: The celebrated visitor center of NASA's Kennedy Space Center is part aerospace museum and part amusement park. The center is organized into mission zones that focus on the past, present, and future of NASA space programs. By far, this is the most expensive attraction in this book. The complex requires parking fees and single-day or multiday entry fees, plus there are optional add-ons. But this remarkably comprehensive center is well worth the expense, and many visitors will spend more than one day on a personal mission to see it all. Throughout the complex, there are thousands of artifacts, dozens of spacecrafts and rockets, informative exhibits, and fun interactive games and simulators. Just a few of the most popular highlights include a complete Saturn V rocket, the space shuttle *Atlantis*, and an impressive rocket park.

The **STARS**: Rising above the complex, an impressive outdoor rocket garden showcases nine different U.S. launch vehicles—a mix of unflown flightworthy rockets and replicas—from different periods in spaceflight. There's a Juno I, which launched Explorer I, the first U.S. satellite, in 1958. A Juno II, which launched early Pioneer missions. A Delta, a 1960s satellite booster, which

24. The Kennedy Space Center Visitor Complex and rocket garden in 2018. Kennedy Space Center Visitor Complex.

notably launched the experimental Project Echo "satelloon," a massive Mylar balloon inflated in orbit to reflect microwave signals.

Other artifacts include a Mercury-Redstone suborbital rocket and the more powerful Mercury-Atlas, which sent the first Americans into orbit. A Gemini-Titan II launch vehicle from Project Gemini. An Atlas-Agena, which launched eight Ranger probes to study the Moon ahead of Apollo. The final flightworthy Saturn IB in existence. A Delta II, the only modern-era rocket in the garden, and a still-active vehicle that launched the famous Mars rovers *Spirit* and *Opportunity* in 2003.

Nearby, the Heroes and Legends zone mixes tributes to members of the Astronaut Hall of Fame with video presentations and space artifacts, including the Sigma 7 capsule from Project Mercury and a suspended Redstone rocket.

For many visitors, a major highlight is the off-site Apollo/Saturn V Center, which involves a bus ride from the main campus departing every fifteen minutes. The ride itself includes a short bus tour of restricted areas at Kennedy Space Center. The exact sites are dictated by current operations but may include the Vehicle Assembly Building and launch complexes like 39A, the pad for the Space Launch System.

Upon arrival to the Apollo/Saturn V Center, multiple exhibits comprise The Race to the Moon zone. Dominating the center is a horizontally suspended Saturn V rocket, made up of both flightworthy and test stages. It's one of only three complete examples remaining in the United States, with the other two displayed at the U.S. Space and Rocket Center in Huntsville AL and Johnson Space Center in Houston. The vehicle is broken apart into stages, and visitors can wander underneath the massive Moon rocket.

Various exhibits present a detailed story about the Apollo program. There are several unflown flightworthy Apollo spacecraft, including a lunar module from a canceled mission and a combined command and service module repurposed for an emergency *Skylab* rescue mission. The Firing Room Theater replays footage from the *Apollo 8* launch above the actual consoles from Apollo mission control, while the Lunar Theater recreates events from the *Apollo 11* lunar landing. Other highlights include the flown *Apollo 14* command module, unenclosed and beautifully illuminated, Alan Shepard's spacesuit, still covered in lunar dust, displays about landing sites, and a chance to touch a Moon rock from *Apollo 17*.

Back at the main visitor complex, a major highlight is the space shuttle *Atlantis*. Outside the entrance, there is a full-scale replica of the shuttle stack, with two solid rocket boosters flanking an orange external tank. Inside, the *Atlantis* orbiter is uniquely displayed at a 43-degree angle. The payload doors are open, allowing a view inside the cargo bay, and the Canadarm is extended, as it might be during orbital missions.

Surrounding exhibits tell the story of the thirty-year shuttle program. There's a memorial to the fourteen astronauts who perished in the *Challenger* and *Columbia* accidents, plus hardware recovered from the wreckage. Nearby, the Hubble Space Telescope Theater has a full-sized replica of the Hubble Space Telescope and a short film about the telescope.

Elsewhere, exhibits focus on the International Space Station and NASA's ongoing exploration of the solar system, including a return to the Moon, asteroid and planetary missions, and the current and future exploration of Mars, along with rover replicas. The latest zone is Gateway: The Deep Space Launch Complex, which focuses on the interstellar travel of tomorrow, including human missions into deep space, a spaceport of the future, and a flown SpaceX Falcon 9 booster.

25. Inside the Apollo Saturn V Center at the Kennedy Space Center Visitor Complex. Kennedy Space Center Visitor Complex.

Included with admission, there are a variety of space-themed IMAX films, with showings available throughout the day. *Mission Status Briefing* is a regularly updated multimedia presentation about current NASA missions.

EVAs: At times, additional Behind the Gates bus tours may be available for an extra fee. These typically require advance reservations, with more information available on the center website. The Cape Canaveral Early Space Tour is about 2.5 hours long and includes historic launch sites around Cape Canaveral and the Air Force Space and Missile Museum, an on-base site with access restricted to tours only.

Cape Canaveral's Rise to Space Tour is an expanded 4.5-hour tour that includes historic launch sites and more. Sites include the Cape Canaveral Lighthouse; a visit inside Hangar C, where restored rockets and other spacecraft are displayed at floor level; Complex 26, the launch site for Explorer 1 in 1958; and a visit to the restricted Air Force Space and Missile Museum.

Countdown: Open daily 9:00 a.m.–5:00 p.m.; April 4–24 9:00–6:00 p.m.

Mission budget: $$$$

Flight team: 855-433-4210 | https://www.kennedyspacecenter.com

Coordinates: 28.5227, -80.6819 | Space Commerce Way, Merritt Island FL 32953

Flight plan: In 2018 the complex opened a new visitor entrance. From I-95 follow signs for Kennedy Space Center. At the intersection of US-1 and FL-405, continue east on Space Commerce Way for about 1.2 miles and turn left into the Kennedy Space Center Visitor Complex.

Rations: The visitor complex has multiple dining establishments, with options including restaurants, fast food stands, ice cream shops, and a coffee/beer/wine bar. Further options can be found in nearby mainland communities.

Sands Space History Center, Cape Canaveral FL

Debriefing: Operated by the U.S. Air Force, this small museum is just outside the south gate of Cape Canaveral Space Force Station. Admission is free, and because it's off-base, the history center can be visited without an access pass by the general public. The history center's on-base parent facility, the USAF Space and Missile Museum, can only be visited on guided tours from the Kennedy Space Center Visitor Complex. The history center's displays include several rocket engines, nose cones, and other small artifacts related to space travel.

The STARS: Artifacts include a Mercury boilerplate capsule used for training recovery forces, a Loki sounding rocket and launcher, and engines from vehicles like the Atlas rocket, a Titan I rocket, and a Mace missile. On display are several nose cones, including one from a Pershing missile and one called the Heatsink, which tested a heat deflection reentry method. Another nose cone is from a Jupiter intermediate range ballistic missile, which were some of the first objects to be launched into space and safely returned, during test flights in the late 1950s. Other exhibits include historic photographs, models, smaller artifacts, and informational displays.

EVAs: Some tours from the Cape Kennedy Visitor Center Complex also visit this museum.

Countdown: Tuesday–Friday 9:00 a.m.–2:00 p.m.; Saturday 9:00 a.m.–5:00 p.m.; Sunday 12:00–4:00 p.m.; closed Monday

Mission budget: Free

Flight team: 321-336-8080 | https://afspacemuseum.org/artifacts/history-center/

Coordinates: 28.4177, -80.6052 | 100 Spaceport Way, Building 90328, Cape Canaveral FL 32920

Flight plan: From I-95, head east on FL-528/A1A/Sea Ray Drive for just over 12 miles. Take the exit for Port Canaveral/North Cargo Piers and continue onto FL-401 for about 2.5 miles. Turn right onto Spaceport Way and follow it to the history center parking lot.

Rations: The history center has a small gift shop but no food. There are many restaurants and stores in nearby mainland communities.

American Space Museum, Titusville FL

Debriefing: Previously called the U.S. Space Walk of Fame Museum, the American Space Museum is a small nonprofit facility focused on missions, astronauts, and the workers behind the scenes who make the U.S. space program possible. Artifacts and exhibits include spacesuits, spacecraft parts, launch consoles, and more.

The STARS: Artifacts include launch consoles from space shuttle and Atlas-Centaur control centers, a space shuttle crew escape suit, an engine piston from NASA's crawler launchpad transporter, and cosmonaut space gloves.

Other artifacts and exhibits are organized by NASA space program. The Mercury gallery has a capsule hatch and a banana pellet of the type fed to NASA's early primate astronauts. The Gemini gallery includes the thruster controller that failed on *Gemini 8*.

The Apollo gallery is the largest, with a focus on each lunar mission plus *Skylab* and the Apollo-Soyuz Test Project. Other exhibits focus on women in the space program and Cape Canaveral. Docent-guided tours are available with an advance reservation required.

EVAs: A few blocks northeast, the waterfront Space View Park has a series of monuments and plaques dedicated to the workers, astronauts, and fatalities from the four NASA human spaceflight programs.

Countdown: Monday–Saturday 10:00 a.m.–5:00 p.m.

Mission budget: $

Flight team: 321-264-0434 | https://spacewalkoffame.org/

Coordinates: 28.6105, -80.8086 | 308 Pine Street, Titusville FL 32796

Flight plan: The museum is in downtown Titusville, a few blocks from **Space View Park**.

Rations: Restaurants can be found throughout the town and surrounding region.

Orbital neighbor: An hour inland from Cape Canaveral, the **Orlando Science Center** is a large interactive museum with a wide range of exhibits, including some topics related to space, like Planet Earth, science fiction, and the future. **Flight team:** www.osc.org.

26. The Moon and Earth as seen from the International Space Station in 2022. NASA.

10   The Rise of Apollo

How to Reach the Moon?

For Apollo, there were two primary challenges above all others. How to launch from Earth and how to land on the Moon. Since its inception, NASA had been designing a massive rocket called Nova, which would be over 370 feet tall and fifty feet wide at its base and use eight F-1 engines. But when von Braun's rocket team joined NASA, they brought along plans for their own heavy-lift rocket, eventually called Saturn V. Compared to Nova, the Saturn V was about ten feet shorter, it was thirty-three feet wide at its base, and it used five F-1 engines. The Saturn V was further along in the development process and deemed more realistic, so NASA ultimately gave von Braun the nod.

Since it would take years to develop the behemoth Saturn V, von Braun's team would first build the smaller and simpler Saturn I, followed by an upgraded Saturn IB, to test Apollo spacecraft during launch and orbital flight. Based upon existing technology, with a first stage that clustered eight Redstone rockets around a central Jupiter core, the Saturn I was still the biggest U.S. rocket to date. Standing about 180 feet in height, depending on the payload, it had a diameter of nearly twenty-two feet at its base. Powered by eight H-1 engines igniting kerosene and liquid oxygen, the rocket had 1.5 million pounds of thrust at launch and was capable of lifting 20,000 pounds into low Earth orbit.

With the launcher plans in progress, the next question was how to land. There were three options. Direct ascent, in which a massive Nova-style rocket, with up to five stages, launched a large spacecraft into orbit. The spacecraft, with three stages, would fly to the Moon, also called translunar injection, where two stages would land. The final stage would lift the spacecraft off the Moon and return it to Earth. The thrust requirements for this approach were enormous, and once the downsized Saturn V was selected, this option was eliminated.

The next option was favored by von Braun and most other NASA officials. Called Earth orbit rendezvous, multiple coordinated launches of Saturn V rockets would place different elements of a lunar mission into orbit. These elements would dock with one another to assemble the lunar spacecraft, which would be fueled above the earth before proceeding to the Moon. Henceforth, the approach was similar to direct ascent, with two stages landing on the lunar surface and one stage returning the spacecraft to Earth. As studies continued, von Braun began to realize this option might take as many as five Saturn V launches.

There was a third option, initially dismissed by all but a few scientists at NASA for being too risky. Called lunar orbit rendezvous, a single large rocket like the Saturn V would put a smaller four-part spacecraft with three astronauts on a course for the Moon. It would be launched in two disconnected pieces to save space. During translunar injection these pieces would dock while traveling over 23,000 mph. Upon reaching lunar orbit, half the craft would stay in orbit—the command and service modules, with one astronaut. Meanwhile, the other half of the craft—the lunar descent and ascent modules—would descend to the surface with two astronauts. After surface exploration, the small ascent module would return to orbit and dock with the command module, reuniting the three astronauts, before being jettisoned. Then the command and service modules would return to Earth, with just the command module splashing down in the ocean.

From a weight and fuel perspective, this compartmentalized plan was the most efficient, allowing each of the four elements to be stripped down to the bare essentials for various tasks. But no one had ever docked two spacecraft in Earth orbit, let alone done it twice on the same mission, hundreds of thousands of miles out in space, where the slightest malfunction could lead to the death of all three astronauts.

As plans proceeded for the Earth orbit rendezvous, it reached a point where only a single voice within NASA was effectively advocating for reevaluation. An engineer named John Houbolt, whose arguments for lunar orbit rendezvous had been repeatedly dismissed, wrote an impassioned nine-page letter to NASA's associate administrator. The current plan was too complex, too costly, and too slow. If the goal was to meet the deadline within the decade, and work within the confines of the budget and capabilities of a single Saturn

27. NASA engineer John Houbolt gives a presentation about lunar orbit rendezvous. NASA.

V, there was no other choice. One by one, the entirety of NASA's leadership
changed their minds, including von Braun. Not only did the advantages of
lunar orbit rendezvous outweigh the risks, but this was the only way to meet
the deadline. The plan was set. Now it was time to start carrying it through.

Apollo Second Thoughts

In September 1962 President Kennedy spent a busy few days inspecting the
various Apollo program elements. He flew to Marshall Space Flight Center
to observe the test firing of a Saturn I first-stage booster. Then he went to
Cape Canaveral to view the launch pads. Next, onward to Houston, where
he visited the Manned Space Flight Center, future home of mission control.
On the morning of September 12, he spoke to a large crowd in the football
stadium at Rice University. "For the eyes of the world now look into space, to
the Moon and to the planets beyond," said Kennedy, going on to describe the
Saturn I and Saturn V rockets. "We choose to go to the Moon in this decade
and do the other things, not because they are easy, but because they are hard."

Kennedy continued to speak about the developing space program during the following year. But as successful Saturn I launches signaled America was gaining footing in the Space Race, and as costs soared, Kennedy began to have doubts. In 1960 the NASA budget had been $401 million, about 0.5 percent of the U.S. budget. The next year, it was $744 million, followed by $1.26 billion in 1962, or about 1.2 percent of the annual budget. In 1963 it went to $2.5 billion, making it 2.3 percent of all expenditures. With the continuation of Apollo, the projections were enormous. The next few years would see increases to $4 billion, $5 billion, and just under $6 billion—the final number representing 4.4 percent of the annual U.S. budget. This is roughly equivalent to $52.5 billion in 2022, about the same as the budget of Maryland.

Then, in mid-September, NASA administrator James Webb delivered to Kennedy more concerning news. The Apollo mission was harder than realized, and there would be no landing during a hypothetical second term should Kennedy win reelection. The soonest would probably be 1969. Disappointed, Kennedy asked if the program was a good idea at all. Webb reassured him the mission was well worth the effort.

But Kennedy seemed to be considering whether loosening the deadline, and reducing associated costs, might be a good idea. Over the past few years, Kennedy had suggested several times that Soviets and Americans might collaborate in space instead of competing. In 1961 he'd asked Khrushchev about the idea at a summit in Austria. The Soviet leader had waffled and ultimately said that disarmament would have to happen first.

Then, on September 20, 1963, Kennedy gave a speech at the UN, where he proposed a joint expedition to the Moon. Why make it a competition? he asked. Why duplicate the research and construction and costs? Observers in Washington DC were stunned at this retreat from Kennedy's Cold War Moon race. Internally, Khrushchev and other Soviets discussed the idea, but little came of it over the following weeks.

On November 21, Kennedy flew to Texas for a weekend visit. Speaking upon arrival at Brooks Air Force Base, he mentioned the Saturn I rocket, scheduled for its first two-stage test launch in December. "While I recognize that there are still areas where we are behind," said Kennedy, "at least in one area, the size of the booster, this year I hope the United States will be ahead."

The next day, President Kennedy was riding in his presidential convertible through a crowd in Dallas. Waiting with a rifle in a sixth-floor window of the Texas School Book Depository was Lee Harvey Oswald. The twenty-four-year-old Oswald was a troubled former U.S. Marine who had previously defected to the Soviet Union before returning home. His first shot hit a curb, but the second and third shots found their target.

"Kennedy is Killed by Sniper" read the *New York Times* headline. The assassination shocked the nation, and memorials were held throughout the world. Five days later, the new president, Lyndon Johnson, delivered a speech to a joint session of Congress titled "Let Us Continue."

"The greatest leader of our time has been struck down by the foulest deed of our time," said Johnson. He outlined his plans to continue the initiatives Kennedy had started, including civil rights legislation, Cold War foreign policy, and "the dream of conquering the vastness of space."

Two days later, Johnson renamed Cape Canaveral as Cape Kennedy—a change reversed ten years later. As vice president, Johnson had chaired the National Space Council, and he was more devoted to space travel than Kennedy, who had viewed the endeavor as a proxy form of Cold War combat. But with JFK's death, the Moon landing would now become a tribute to the fallen president, and few would seriously challenge its expenditures. When Johnson submitted the next U.S. budget, he made cuts across the board with one exception, an increase for NASA.

Saturns, Geminis, and Voskhods

In January 1964 the two-stage test of the Saturn I proceeded flawlessly. For the first time, a live Saturn IV-B—the second stage, which burned liquid hydrogen and liquid oxygen—was launched. America's first heavy-lift rocket was complete. Finally, the Americans had exceeded the capabilities of the chief designer's R-7.

In the Soviet Union, Korolev had no plans to go quietly. He'd been following the developments of the Gemini and Apollo programs closely. A new Soviet spacecraft called Soyuz was in development, which would be the Soviets' solution for both orbital and lunar missions.

The orbital variant of the Soyuz would use an innovative three-module design. At the front was a nearly spherical orbital module with a workspace and docking unit. In the middle was a bell-shaped descent module with

28. The first Saturn I arrives at Kennedy Space Center in 1961. NASA.

crew seating for two or three and a heat shield for landing. And at the rear was a cylindrical service module with solar panels, thrusters, and propulsion rockets. For lunar missions, additional Soyuz units, including a translunar rocket module and a fuel tanker, would dock to the crewed orbital craft. But all of this was still years away from a first launch.

Concerned about losing the lead in the Space Race, Korolev decided to hastily modify the small Vostok capsule into two new designs called Voskhod. Using the same spherical reentry vehicle, one variant crammed three cosmonauts inside—however, due to limited interior volume, they couldn't wear spacesuits. The other version allowed for two cosmonauts in spacesuits, plus an airlock. Lacking the space for ejector seats, the Voskhod would use a retrorocket to slow its descent over land. When the first test drop from 33,000 feet ended in a crash, Korolev still forged ahead.

In April 1964 NASA launched an unmanned Gemini spacecraft on an orbital test flight. It was basically an advanced and enlarged Mercury capsule, painted black and designed to hold two astronauts. Behind the heat shield, there were two attached cylindrical units, painted white. The retrograde module carried fuel and a retrograde rocket for returning to Earth. The equipment module carried fuel cells, batteries, and drinking water.

With sixteen attitude and maneuvering thrusters and a cockpit resembling that of a fighter jet, Gemini was built to truly fly in space—the first spacecraft that could change its orbit and speed. It would need these capabilities to practice the orbital techniques, like rendezvous and docking, that would be vital for Apollo. Given this maneuverability, early designs even toyed with deploying a glider in the atmosphere so Gemini could fly back to the runway like a plane. But this type of glider, called a Rogallo Wing, had deployment issues, and it was abandoned in favor of the now-typical splashdown.

In early October 1964, a successful Soviet test launch paved the way for *Voskhod 1*. In addition to a cosmonaut pilot, Korolev sent up an engineer and medical specialist. The three-person flight set many records, including first multiperson crew, first nonpilots, and first spaceflight without spacesuits. Although it was an update to the previous Vostok model, cleverly worded statements by the Soviet news agency that announced *Voskhod 1* made it sound like the arrival of an advanced new vehicle.

29. Ed White becomes the first U.S. space walker, and the world's second, on *Gemini 4* in 1965. NASA.

11   Adventures and Lessons

How the First Spacewalk Ended in a Ski Trip

In mid-May 1965 *Voskhod 2* launched from the Baikonur Cosmodrome. During the second orbit, the commander, Pavel Belyayev, deployed the Volga airlock, an eight-foot-long inflatable tube. Cosmonaut Alexei Leonov squirmed his way into a backpack oxygen tank. After Leonov crawled into the airlock, Belyayev closed the hatch and drained the air. A minute later, Leonov floated out into space, attached to a 16-foot tether. The first spacewalker in history.

During his ten minutes outside, problems arose for Leonov. The suit was incredibly hot, and Leonov claimed he began sweating profusely. The arms and legs of his spacesuit over-inflated from air pressure, which limited his mobility. To reenter the airlock, he had to open the bleed valve on his spacesuit, risking decompression sickness.

Once aboard, problems continued when the automatic reentry rocket failed. Belyayev had to manually fire the backup, which caused the spacecraft to come down hundreds of miles off course. Parachutes brought them down in the Perm region, near the Ural Mountains. The capsule landed on deep snow between two fir trees in dense forest.

Helicopters located them within hours, but finding no place to land, all they could do was drop supplies and warm clothes. The cosmonauts spent a below-freezing night in the hatchless capsule, claiming to hear howling wolves. The next afternoon a rescue party on skis arrived, but given the hour, it was too late to egress. They all built a large camp with a fire and were forced to spend another night in the woods. The next morning, the historic mission took its final turn, with the cosmonauts donning boots and clutching poles. As they departed the capsule, the first human spacewalk ended with a ski trip.

In Moscow, the new Soviet leader was less than impressed. During the same day *Voskhod 1* had spent in orbit, Nikita Khrushchev had been ousted in a coup, allegedly for his increasingly erratic behavior. Replacing him was

Leonid Brezhnev, a more pragmatic leader, less smitten with expensive space spectacles. From a political standpoint, Korolev had lost one of his most enthusiastic supporters.

Peculiarities of Orbital Mechanics

Back in the United States, Gemini had a tough act to follow. In March, the first manned Gemini mission launched on a 101-foot Titan II rocket. Derived from an U.S. Air Force ICBM (intercontinental ballistic missile), this two-stage liquid-fueled rocket used hypergolic propellants, hydrazine and nitrogen tetroxide, which spontaneously ignited upon contact.

During three orbits, Gus Grissom and John Young tested the new U.S. spacecraft. Using thrusters, they executed the first orbital maneuver in a crewed spaceflight by shortening their orbit to make it more circular. This first flight was a short and sweet success, but upcoming missions would soon discover new challenges.

In June, *Gemini 4* followed with an ambitious mission lasting over four days. Early in the first of sixty-two orbits, James McDivitt turned the spacecraft around. The goal was a seemingly simple rendezvous with the spent Titan second stage, which had boosted them into orbit and now trailed behind. But when McDivitt thrusted toward this target, *Gemini 4* moved downward and away.

For a trained jet pilot, this seemed nonsensical. Each time he tried to thrust toward the Titan stage, the distance *increased*. After several attempts, the barrel-shaped rocket was maybe a half mile above and behind them. That said, without a radar instrument onboard, the two astronauts couldn't even agree on the distance. With half their thruster fuel spent, they wisely abandoned the experiment.

Without realizing it, *Gemini 4* had stumbled across a key component of orbital mechanics. When *Gemini 4* fired its thrusters opposite the direction of travel, called retrograde thrusting, they ever so slightly decreased their orbital velocity. This allowed Earth's gravity to pull them into a lower, and therefore shorter, orbit. Since the spacecraft now had less distance to travel around the earth, *Gemini 4* moved relatively faster away from the spent Titan stage, which was moving in a higher and therefore larger, and relatively slower, orbit. It was a counterintuitive principle, and a concept NASA astronauts would have to master before accomplishing their plans for Apollo.

30. The *Gemini 7* spacecraft as seen from the *Gemini 6A* spacecraft during the world's first rendezvous in space. NASA.

During their third orbit, crewman Ed White made the first American space-walk. This EVA (extravehicular activity) was much simpler than Leonov's. White opened the hatch above his head, stood up, and floated out—attached to a tether. White carried a pressure gun that allowed him to twirl around above the globe. The photos and videos were exceptional, especially when compared to the poor-quality imagery from *Voskhod 2*.

Project Gemini and Preparing for Apollo

The Gemini flights came at a rapid pace, henceforth. In August, *Gemini 5* orbited the earth for eight days, a new world record, which matched the amount of time it would take to fly to the Moon. In October, Wally Schirra and Ed Stafford waited on the launchpad in *Gemini 6*. From a nearby pad, an Atlas rocket was launched with a specially adapted Agena Target Vehicle—basically an uncrewed spacecraft built for docking and rendezvous. Six minutes into launch, the Agena Target Vehicle exploded, and *Gemini 6* was postponed.

In early December, *Gemini 7* was launched, with Frank Borman and Jim Lovell, on a record fourteen-day mission. This would almost double the time previously spent in space by *Gemini 5*. Eleven days later, the marathon mission received an exciting jolt when the rescheduled *Gemini 6A* flew up to meet them. Following the lessons learned about orbital mechanics, *Gemini 6A* made a series of course corrections, using a lower orbit to catch up to *Gemini 7*. Then they maneuvered upward and came within a few feet of the other craft—the first active rendezvous in space.

Gemini 8 launched on March 16, the fortieth anniversary of Robert Goddard's first liquid-fueled rocket flight. This marked the first time an American civilian had flown in orbit: Neil Armstrong, who was joined by David Scott. The goal was to rendezvous with an Agena Target Vehicle and conduct the first space docking in history.

The radar acquired the Agena at 206 miles out, and after sighting the target craft, they turned flight control over to the guidance computer. After a visual inspection, Armstrong piloted *Gemini 8*'s nosecone slowly into the Agena's docking adapter. The latches clicked and a green indicator lit up. "Flight, we are docked," radioed Scott, calling it a real smoothie.

But a moment later, Scott noticed the combined spacecraft was rolling to the right. Armstrong fired the orbit attitude and maneuvering system (OAMS) thrusters, which briefly stopped the roll. But then it started rolling again. Soon the OAMS thruster fuel gauge had dropped to 30 percent. While Armstrong struggled to stabilize the vehicle, Scott undocked from the Agena. As Armstrong backed *Gemini 8* away, the spin worsened until they were tumbling almost one full rotation every second. Armstrong shut down the OAMS system and successfully used the reentry control thrusters in the nosecone to stop the roll. After testing the OAMS system, they discovered one of the thrusters had jammed in the on position. Now having only 75 percent of their reentry thruster fuel, the planned three-day mission had to be aborted after only ten hours in orbit.

Troubles continued for *Gemini 9*, when the original crew of Elliot See and Charles Bassett were killed in a T-38 crash in St. Louis. The backup crew of Ed Stafford and Gene Cernan were elevated to the prime crew. But once again, the Agena Target Vehicle exploded during launch. A backup augmented target docking adapter was lifted into orbit, but when *Gemini 9A* rose to meet it, the crewmen saw that the nosecone fairing was still attached. During a two-hour spacewalk, designed to test abilities to work in space, Cernan discovered the lack of foot and hand holds on the spacecraft made maneuvering very challenging. Overheated and operating under limited visibility due to a fogged visor, Cernan returned to the cabin. Clearly, an update to the Apollo spacesuits would be needed.

Project Gemini needed to rebound, and it did so over the final three flights. During *Gemini 10*, Michael Collins and John Young docked with a successfully launched Agena Target Vehicle, using the onboard booster to climb to a higher orbit, then rendezvoused with the abandoned Agena from *Gemini 8*. Collins spacewalked over to the dormant vehicle and retrieved a micrometeorite collector.

On *Gemini 11*, Pete Conrad and Richard Gordon also successfully docked with an Agena and used it to boost themselves to a record orbital altitude of 853 miles. After connecting a tether to the Agena, they undocked and spun the two spacecraft around their center of mass, creating a weak force of artificial gravity, just like the spinning space station von Braun had once envisioned.

The final mission was *Gemini 12*, with Buzz Aldrin and Jim Lovell running through final techniques needed for Apollo, including spacewalking. The spacecraft had been updated with external handles, and ground training had expanded to include underwater training. During the course of three spacewalks, Aldrin proved that astronauts could successfully work outside a spacecraft. As Project Gemini came to an end in November 1966, all its objectives were complete.

31. A Soyuz spacecraft in orbit in 1975. NASA.

12 A Focus on the Moon

The Soviet Lunar Program

Throughout 1965, as Project Gemini was pushing American space capabilities into new territory, Korolev was lobbying Brezhnev to officially approve a Moon project. Since the early sixties, the Soviet government had made vague proclamations about landing a cosmonaut on the Moon, but little progress had been made.

For years, Korolev had been dreaming big—not only a mission to the Moon but crewed flybys of Mars and Venus as well. Toward those goals, Korolev began designing the N-1, a massive conical rocket that would stand 344 feet tall and be 56 feet wide at its base. Exceeding even the capabilities of the mighty Saturn V, the N-1 first stage would have thirty engines burning kerosene and liquid oxygen, producing over ten million pounds of thrust during launch. The result would be an ability to launch almost a hundred tons into low Earth orbit or put over twenty tons on a trajectory to the Moon.

The rest of Korolev's plan involved a modified Soyuz spacecraft carrying two cosmonauts to lunar orbit, and a one-man lander called Lunniy Korabl. The small Lunniy Korabl included a spherical cosmonaut cabin atop a barrel-shaped instrumentation unit with four landing legs.

As NASA's Apollo program made major public strides toward Kennedy's within-the-decade challenge, the Soviets coyly made official statements denying any interest in a Moon landing. But internally, the Soviet government was hoping to land the first cosmonaut on the Moon in 1968, and once again beat the Americans.

Mysteries of the Moon

Once the race to land a man on the Moon had begun, along came the realization of how little anyone actually knew about the target. Telescopes had long showed the Moon was covered in craters. But were these impact features,

32. The crater Daedalus on the Moon's far side, as seen by Michael Collins during *Apollo 11*. NASA.

collapsed volcanic calderas, or both? If they were volcanic, might the Moon be seismically active, meaning a landing could experience moonquakes? If they were impact features, it was possible the Moon might be covered by a deep layer of loose ejecta. And if a spacecraft tried to land on this layer of pulverized dust, would it sink underneath as if it were in quicksand?

Beyond such practical considerations about landing conditions, there were theoretical questions too. First and foremost, how had the Moon formed? By the early 1960s, there were three competing theories. One theory said the Moon had formed elsewhere in the solar system and was captured by Earth's gravity. Another theory said the Moon formed from the same protoplanetary disk of coalescing dust as the earth. A third theory, proposed by astronomer George Darwin, son of evolutionary theorist Charles Darwin, suggested centrifugal forces had spun the Moon off from an early molten Earth, leaving the Pacific Ocean behind as a scar.

Each theory had its adherents, but none of the theories could account for several anomalies of the Earth-Moon system. To start, the Moon is quite large, about a quarter the size of Earth, which does not have enough gravity to capture it. The earth has never rotated fast enough to squirt out a primordial molten moon. And if the earth and Moon had accreted at the same time, from the same swirling disk, then why does the Moon's orbital angular momentum exceed the rotational angular momentum of the earth? In all other planet-moon systems in the solar system, this relationship is reversed. Certainly, something odd must have happened in the formation of Earth's lone natural satellite.

From Tranquility to Anything But

At NASA, the period before Apollo's manned missions began, in early 1967, was spent trying to answer some of the most pressing lunar questions to support a safe landing. It was 1961 when NASA began launching Ranger spacecrafts toward the Moon, but most early efforts were met with failure. It wasn't until *Rangers 7, 8,* and *9* crashed into the lunar surface, with cameras snapping until impact, that the project began returning high-quality photos. Some of these images revealed craters so small they had to have been formed by meteors. Still, the solidity of the surface remained in question.

The Soviets had accomplished the first soft landing on the Moon with Korolev's *Luna 9* in February 1966, which demonstrated the surface could support a spacecraft. Then, in early June of that year, NASA accomplished their own soft landing when *Surveyor 1* touched down in the Ocean of Storms—one of the vast lunar maria, or dark plains that resemble seas—on the western portion of the Moon's near side.

Over the following two years, NASA would send six more Surveyor probes to the lunar surface. Next up for NASA were lunar orbiters, which mapped the surface looking for landing sites. Five orbiters succeeded in mapping 99 percent of the Moon at a resolution of 200 feet or better. Upon analysis, signs seemed to favor the Sea of Tranquility, where there appeared to be some flat surfaces, free from craters and debris, upon which a spacecraft might land.

But as Apollo progressed toward its goal, back on Earth—and particularly in the United States—things were anything but tranquil. Since Kennedy's assassination, a series of counterculture revolutions had been erupting around the country. Some protests related to racial discrimination and civil rights. Others were focused on gender inequity and women's rights. Peace movements reflected the seemingly endless conflicts of the Cold War. Since reelection in 1964, Johnson had rapidly escalated U.S. involvement in the Vietnam War. While NASA was marching toward the Moon, increasing numbers of protestors were marching across America.

Sites in the South and on the Gulf Coast
Eight Adventures and Seven Orbital Neighbors

Celestial overview: Spanning the South and Gulf Coast, this chapter covers the largest geographic region in this book, organized here from east to west. Adventures range across North Carolina, Georgia, Alabama, the Florida Panhandle, Mississippi, and Texas.

The main highlights of this region include two must-see sites for space enthusiasts about 750 miles apart: Space Center Houston and the U.S. Space and Rocket Center in Huntsville AL. Both facilities serve up amazing spacecraft, rockets, exhibits, and experiences. Located midway between these two facilities, the Infinity Science Center and NASA Stennis Visitor Center in Mississippi is another appealing site with a smaller but worthy exhibition.

The remaining adventures and orbital neighbors include several large aviation-focused facilities and science museums with space artifacts and exhibits, plus a series of astronomy observatories, and more. The lone outlier—both in nature and geography—is launch viewing from South Padre Island near the SpaceX Starbase.

Many of these places are far apart, and most space enthusiasts will likely plan these adventures during short trips to individual sites. Still, various custom itineraries are possible and, in some cases, can include highlights from other regions. For example, a road trip down the middle of the country might include some combination of Strategic Air Command in Nebraska, the Cosmosphere in Kansas, Stafford Air and Space plus other museums in Oklahoma, and Space Center Houston and/or other Texas options. An East Coast road trip might combine New York City with Washington DC and Virginia, with a stop in North Carolina on the way to Cape Canaveral in Florida. A southwestern road trip might add some Texas sites to those in New Mexico and Arizona. Yes, the distances are far, but—when considering astronauts traveled 240,000 miles to the Moon—not impossible.

Museum of Life and Science, Durham NC

Debriefing: This 84-acre campus combines a two-story museum with a large outdoor park. Outside offerings include nature play structures, treehouses, remote-control sailboats, a train ride, a dinosaur trail with life-sized dinosaur models, an animal farm, and a small zoo with endangered red wolves, bears, and more. Indoor exhibits include a butterfly house, an insectarium, several interactive exhibits, and a small zoo with birds and reptiles. The highlight for space enthusiasts is an aerospace gallery with a mix of space-flown artifacts, replicas, and displays.

The STARS: Outside the museum, an authentic Redstone missile has been painted to simulate the Mercury-Redstone rocket, complete with a mock-up capsule. Inside, highlights include the space-flown Mercury-Atlas 5 capsule, which successfully launched Enos, the first chimpanzee and third primate, into orbit and returned him safely to Earth.

The space highlights focus on *Apollo 15*, featuring a re-creation of the lunar landing site. A training mock-up of the lunar module is surrounded by two replica astronauts simulating events from the mission, with commander David Scott performing his feather and hammer experiment. During the experiment, Scott dropped both objects together, which fell in unison and impacted the lunar surface simultaneously. This confirmed Galileo's hypothesis that, in a vacuum, objects fall at the same rate regardless of their mass.

Other Apollo artifacts include a lunar rover front wheel assembly, showing the unique woven-wire tires, plus several rover prototype models. Two space-flown objects are Neil Armstrong's bio-belt assembly and Buzz Aldrin's dosimeter from *Apollo 11*. A command module engineering vehicle, used for electrical system testing, can be entered to get a sense of the interior size of the spacecraft. On display is a J-2 engine from the Saturn V rocket, plus an RL-10 engine from an Atlas-Centaur.

There are several spacesuits from the *Skylab* missions, an inner suit and two spacesuits, one of which was worn during a spacewalk on *Skylab 4*. A Hasselblad camera from Project Gemini. A display on space food includes examples from Mercury, Gemini, Apollo, and a space shuttle. An interactive exhibit called Launch Lab focuses on the properties of aerodynamics, while Tinker Lab involves circuits, gravity, and more.

EVAs: When visiting the museum, allow enough time to explore the large campus and outdoor exhibits.

Countdown: Open daily 10:00 a.m.–5:00 p.m.

Mission budget: $$

Flight team: 919-220-5429 | www.lifeandscience.org

Coordinates: 36.0293, -78.8989 | 433 West Murray Avenue, Durham NC 27704

Flight plan: The museum is just north of I-85, using Exit 175, in Durham NC

Rations: The museum has the **Sprout Cafe** and the **Elements Coffee Bar** on-site. More restaurant options are available throughout the Durham area.

Orbital neighbors:

Aviation enthusiasts may be interested in a side trip to the Outer Banks, where they can stop by the **Wright Brothers National Memorial**, a National Park Service unit at Kitty Hawk, site of the historic 1903 flight.

In Wilmington NC the **Cape Fear Museum of History and Science** includes an interactive learning center for kids called Space Place, based upon the International Space Station. **Flight team:** www.capefearmuseum.com.

If you want to toss in a quirky hop out of the car, you can see Moon boot prints in Myrtle Beach SC. On April 21, 1972, Charlie Duke became the tenth person to walk on the Moon. A year later, Duke became the first astronaut wearing Moon boots to step in wet cement at the **Myrtle Beach Convention Center**, when he was inducted into the South Carolina Hall of Fame. You can visit the boot prints, by the flagpoles, at 2101 North Oak Street, Myrtle Beach SC. **Flight team:** www.visitmyrtlebeach.com.

Located on campus at the University of South Carolina Aiken, the **DuPont Planetarium** offers astronomy shows and films in a domed theater. There's also a telescope observatory for public viewing.: www.usca.edu/rpsec/departments/planetarium.

Tellus Science Museum, Cartersville GA

Debriefing: Located an hour northwest of Atlanta, the Tellus Science Museum has four permanent galleries and three rotating galleries. Exhibits focus on minerals, fossils, archaeological artifacts, and an interactive invention playground for children. Outside there's a solar demonstration home built by West Virginia University. Space enthusiasts will be most interested in the Science in Motion gallery, which uses artifacts and exhibits to tell the story of a century of evolution in transportation technology. Examples range from an 1896 Ford quadricycle to a replica *Wright Flyer* to a Bell-47, the first civilian helicopter, plus a variety of space-related items. The museum also has a planetarium and telescope observatory.

A unique replica of the Apollo 1 capsule uses cutaway panels to show the interior of the module damaged by fire during the 1967 launchpad disaster. Other items include a Soviet Sokol KM rescue suit for a 1970s Soyuz spacecraft and a TZK jumpsuit from the 1980s. There's a NASA space shuttle fuel cell and a pair of tires. Models depict key rockets of the space age, including the Saturn IB, Saturn V, Soviet Proton, Soyuz R-7, and a space shuttle.

The Bentley Planetarium offers a variety of shows on a domed projection screen throughout the day. Tickets are required for a small extra fee, and schedules are available on the museum website.

EVAs: The Tellus Observatory has a 20-inch telescope offering visitors views of the Moon and planets during special events and astronomy workshops, with more information available on the museum website events page.

Countdown: Open daily 10:00 a.m.–5:00 p.m.; closed major holidays

Mission budget: $$

Flight team: 770-606-5700 | https://tellusmuseum.org

Coordinates: 34.2421, -84.7703 | 100 Tellus Drive, Cartersville GA 30120

Flight plan: The museum is located just off I-75 near the junction with US-411.

Rations: There's no food at the museum, but there are options in nearby communities.

Orbital neighbors: Located in Atlanta, the **Fernbank Science Center** is a museum operated by the DeKalb County School System. In addition to general science and nature topics, the center has the authentic command module that flew in space on the uncrewed *Apollo 6* test flight of the Saturn V rocket. There is also a planetarium and dome theater offering star shows and films, plus a telescope observatory offering public viewings on certain nights. **Flight team:** www.fernbank.edu.

U.S. Space and Rocket Center, Huntsville AL

Debriefing: The Huntsville area is rich with spaceflight history, and the U.S. Space and Rocket Center is a must-see for space enthusiasts, especially families with kids. The center is basically a space museum mixed with a small amusement park, and families will find a welcome blend of education and entertainment. In addition to many space artifacts on permanent display, there's also a featured exhibits gallery with rotating exhibitions on space and related technology. The center is home to Space Camp and Space Academies, which offer themed educational programs for kids in science, technology, and more. This is the official NASA visitor center of the Marshall Space Flight Center, the U.S. government's civilian rocketry and spacecraft propulsion research center. Marshall Space Flight Center was where the Saturn rockets were developed, and work today focuses on projects like the Space Launch System. The U.S. Space and Rocket Center is near the Redstone Arsenal, a U.S. Army post that includes the army's Aviation and Missile Command and the future location of U.S. Space Command.

The STARS: There's no shortage of highlights here. The first thing arriving visitors will notice are several outside artifacts rising above the center, including a replica Saturn V and a test article Saturn I. Elsewhere on the outdoor campus, there's a full-size mock-up of the complete space shuttle stack. This lofted exhibit includes an external fuel tank, two solid rocket boosters, and the mock-up Pathfinder orbiter that was used for NASA facilities testing in the early 1980s. A comprehensive restoration began in 2021 and should be complete by 2024.

Other outside artifacts include a shuttle training aircraft, NASA 945, which simulated the flight and landing characteristics of the orbiter. There's a Talon

T-38 trainer, a Chinook helicopter, and an A-12 Oxcart. Undergoing renovation at time of research, an updated rocket park will likely include an Atlas, Titan, and other launch vehicles.

In the Saturn V Hall, literally the biggest highlight is a complete test article Saturn V, which is lofted horizontally and separated into individual stages, displayed sequentially. This is one of only three remaining complete Saturn V rockets, with the other two located at Kennedy Space Center Visitor Complex and Space Center Houston. Also in this massive exhibit hall is the space-flown *Apollo 16* command module. Mock-ups of the lunar module and the lunar rover. A Moon rock from *Apollo 12*. The authentic mobile quarantine facility from *Apollo 12*, which visitors can partially enter.

Another highlight of the gallery is a full-size *Skylab* training module, which can be entered by visitors. This is one of only two full-size training units, with the other on display at Space Center Houston. Meanwhile, the flightworthy backup is on display at the National Air & Space Museum in DC. Nearby, a unique item is the largest surviving fragment of the space-flown *Skylab*—the

33. A replica Saturn V towers over the U.S. Space & Rocket Center, while inside, an authentic Saturn V is displayed horizontally. Mike Bezemek.

oxygen tank. After the first U.S. space station's orbit decayed, *Skylab* broke apart during reentry with some pieces crash landing in Australia in 1979.

Other items include a V-2 rocket, a miniaturized model of Werner von Braun's office at Marshall Space Flight Center, and plenty more. The exhibit called Space Craze displays items of pop culture related to the space age, including toys, games, models, and media. ISS: Science on Orbit combines walk-through mock-up modules of the International Space Station with displays about eating, sleeping, living, and working in space. The SparkLab is an interactive design space for problem solving and invention. And a series of temporary galleries hosts rotating exhibitions on relevant topics—check the website for current offerings.

EVAS: The center has a variety of rides, activities, and experiences. Rides like the Moon Shot drop tower and G-Force Accelerator are included with general admission. Other activities require an additional fee, including the multi-axis trainer, the Apollo 11 Virtual Reality Experience, the Discovery Shuttle simulator, and more. Water activities include a virtual reality snorkeling

34. The space-flown command module from *Apollo 16*. Mike Bezemek.

experience and the Underwater Astronaut Trainer, including scuba diving and guided underwater walking for non-swimmers. The Intuitive Planetarium offers hosted star shows and a variety of space-themed films. A bus tour of Marshall Space Flight Center was suspended due to COVID-19 but may resume in the future. Visit the center website for current offerings, advance tickets, and more information.

Countdown: Monday–Friday 10:00 a.m.–4:00 p.m.; Saturday 10:00 a.m.–5:00 p.m.; Sunday 11:00 a.m.–5:00 p.m.; closed major holidays

Mission budget: $$$

Flight team: 800-637-7223 | www.rocketcenter.com

Coordinates: 34.7113, -86.6532 | One Tranquility Base, Huntsville AL 35805

Flight plan: The center is off I-585 a few miles southwest of downtown Huntsville.

Coordinates: 34.9545, -86.892 | 28232 Upper Elkton Road, Elkmont AL 35620

Rations: The center has an on-site restaurant: the **Mars Grill**. More restaurants abound throughout the Huntsville area.

Orbital neighbors: At the **Ardmore Welcome Center** in Elkmont AL, near the Tennessee state line, 29 miles northwest of the U.S. Space and Rocket Center, stands a surplus Saturn IB rocket.

National Naval Aviation Museum, Pensacola FL

Mission parameters–Base access: Since a terrorist shooting killed three people in 2019, access to Naval Air Station Pensacola has been restricted to Department of Defense ID cardholders and veterans with a Veteran Health ID card. Civilians cannot access the base, including the museum, unless escorted by a Department of Defense ID cardholder as part of the Trusted Travel program. Visit the museum website for current information.

Debriefing: The focus at the National Naval Aviation Museum is understandably on aviation, with an impressive display of over 150 restored aircraft. But there are also several excellent space artifacts and a virtual reality experience related to the Apollo and Skylab programs. In the aviation exhibit, visitors will find a replica A-1 Triad, the U.S. Navy's first airplane, built in 1911. In addition to the many airplanes, there are also several blimp control cars. Other exhibits include dioramas depicting artifacts and scenes from World War I and World War II. The museum is at Naval Air Station Pensacola, the home base of the Blue Angels, and four characteristic Blue and Gold A-4 Skyhawks are on display.

The STARS: Among the highlights of the museum's space exhibits is the command module from *Skylab 2*, which carried three astronauts with U.S. Navy backgrounds—Pete Conrad, Paul Weitz, and Joseph Kerwin—to the *Skylab* space station for a twenty-eight-day mission.

Elsewhere, there's a replica lunar module and a boilerplate Apollo capsule used by the navy to practice retrieval at sea. Other artifacts include a spacesuit used by Jim Lovell, a chronograph used by Gene Cernan, and more. Some displays discuss astronaut training at the base.

For a fee, the Apollo 11 Virtual Reality Experience allows the audience to launch atop a Saturn V, land on the Moon, and return to Earth—complete with seats that vibrate, pitch, and heat up upon reentry.

Countdown: Open daily 9:00 a.m.–4:00 p.m.

Mission budget: Free

Flight team: 850-452-8450 | www.navalaviationmuseum.org

Coordinates: 30.3488, -87.3038 | 1750 Radford Boulevard, Pensacola FL 32508

Flight plan: The museum is located inside **Naval Air Station Pensacola**, a few miles southwest of downtown Pensacola.

Rations: The nearby city of Pensacola has many restaurants.

Infinity Science Center and NASA Stennis Visitor Center, Pearlington MS

Debriefing: Located on the Pearl River in southwestern Mississippi, the John C. Stennis Space Center is NASA's primary rocket engine testing facility. Three massive test stands were built in the 1960s to test Saturn V engines like the F-1 and J-2, which had to be transported by barge. Later tests involved engines from the space shuttle program and, more recently, the Space Launch System. While the space center can no longer be toured by the public, the nearby Infinity Science Center and NASA Stennis Visitor Center is a 72,000-square-foot facility combining space artifacts, related exhibits, and activities. Other exhibits explore the ocean, carnivorous plants, and hurricane prediction.

The STARS: The largest highlight of the museum's outdoor exhibition is the authentic S-IC first stage of the Saturn V rocket from the canceled *Apollo 19* mission. Nearby, an Apollo lunar module replica and some stand-alone engines allow closer inspection. These include an F-1 from the Saturn V first stage and an H-1 from the Saturn S-1B first stage.

Inside the center's Earth and Space Gallery, you'll find the space-flown *Apollo 4* command module from the first uncrewed test flight of the Saturn V rocket. A cutaway scale model shows how the astronauts sat for launch and descent. There's also a lunar module trainer, an *Apollo 15* Moon rock, and several spacesuits, including Mississippi native Fred Haise's full suit from *Apollo 13*, a training suit from Neil Armstrong, and an A7L training suit.

A mock-up of the Destiny module from the International Space Station allows walk-throughs, plus experiment stations and a crew sleep station replica. Another display includes components from the space shuttle's RS-25 main engine. Several exhibits relate to NASA's current programs, like the

Space Launch System and the *Orion* spacecraft, including a mock-up capsule that can be entered. There's an innovative aerospike engine, and rounding out the displays are further spacecraft parts, engines, and models.

EVAs: The center's 3D theater offers movies throughout the day on a wide range of topics, including space travel. A variety of educational programs is available for kids. While bus tours of Stennis Space Center no longer occur, a virtual tour is possible. An Apollo history guided walking tour is also available.

Countdown: Thursday–Sunday 10:00 a.m.–6:00 p.m.

Mission budget: $$

Flight team: 228-533-9025 | www.visitinfinity.com

Coordinates: 30.3113, -89.6048 | One Discovery Circle, Pearlington MS 39572

Flight pean: Take Exit 2 South off I-10 on the Mississippi Gulf Coast.

Rations: The **Infinity Cafe** is on-site, with more options in nearby communities.

Space Center Houston, Houston TX

Debriefing: They say things are bigger in Texas. Space Center Houston certainly complies, with a 250,000-square-foot complex offering a mix of space education and entertainment, attracting over a million students, teachers, and other visitors annually. It is the official NASA visitor center of Johnson Space Center. A popular tram tour allows you to visit the historic Apollo mission control and the astronaut training facility. You can also go inside a space shuttle replica mounted atop a modified 747.

The STARS: An impressive exhibition awaits here. Upon arrival, visitors will be greeted by the space shuttle *Independence* replica. It sits atop the original NASA 905 shuttle carrier aircraft—one of two modified Boeing 747s used to transport the shuttle orbiters between NASA facilities. Visitors can explore the interiors of both the shuttle carrier aircraft and the *Independence* orbiter, where they'll find further exhibits about the shuttle program.

The most popular attraction at the center is the NASA tram tour, included with general admission, which takes passengers to a series of historic Johnson Space Center locations. One stop is the astronaut training facility, which has

35. One of the most recent highlights at Space Center Houston is a space-flown and landed SpaceX Falcon 9 booster. Space Center Houston.

supported NASA missions since 1975. These days, the trainers simulate current endeavors like the International Space Station, familiarizing crew members with the layout before their mission, and the Orion capsule. Another stop is the historic mission control for the Gemini and Apollo programs. Finally, the rocket park includes a hangar with a complete Saturn V rocket, composed of three flight-certified stages and the command/service module from the canceled *Apollo 19* mission. Outside rockets include a Mercury-Redstone, a Gemini-Titan, and a Little Joe II.

More highlights include several space-flown vehicles: the Faith 7 capsule from *Mercury 9* in 1963, the Gemini 5 capsule from 1965, and the *Apollo 17* command module America, which orbited the Moon a record seventy-five times in 1971. Among a display of Moon rocks there is one sample that can be touched. Other authentic items include the podium from John F. Kennedy's second Moon challenge speech at Rice University in 1962. Plus the modified lithium hydroxide scrubber canister hastily prototyped on the ground during *Apollo 13*.

Elsewhere, there's a lunar module test article, a lunar rover trainer, a *Skylab* trainer, and a docking module trainer for the *Apollo-Soyuz* joint mission between the U.S. and U.S.S.R. A series of historical replicas includes Robert Goddard's first liquid-fueled rocket from 1926 and a full-scale replica of Explorer 1, the first U.S. satellite in orbit in 1958. Items from the space shuttle era include an orange launch and entry suit and a test ejection seat like those used on the first four shuttle test flights.

The International Space Station (ISS) gallery mixes space-flown artifacts, like astronaut food, with related items like mock-up station materials subjected to high-speed debris impact tests. Scale models depict the ISS, a space shuttle, and more. Other displays focus on NASA's humanoid Robonaut 2, some spacesuits, and the water purification system. Interactive exhibits relate to robotics, while mission briefing officers present live shows about living on the station.

Mission Mars is an interactive exhibit where you can climb inside a replica Orion capsule, touch an authentic Mars meteorite that landed on Earth, and witness a simulated Martian sunset. Other displays examine the history of Mars exploration and the anticipated hardware needed for reaching and living on the Red Planet.

A recent addition to the outdoor exhibition is a space-flown and landed SpaceX Falcon 9 booster, one of only two on display at time of research, with the other located at the Kennedy Space Center Visitor Complex at Cape Canaveral FL. The Falcon 9 is lofted horizontally, and visitors can inspect, from underneath, the nine Merlin engines, landing legs, reentry grid fins, and fuselage.

The center has several temporary galleries with rotating exhibitions. Visit the website for current offerings.

EVAs: The center offers a variety of live presentations, included with general admission, that run about 15–25 minutes: the Mission Briefing Center offers live presentations about current missions; New Perspectives is a live presentation about life aboard the International Space Station (ISS); BOOM is a live presentation about the science behind rocket combustion. Elsewhere, various films are available. The Destiny Theater presents a space documentary about NASA history. Located in the five-story 4K Space Center Theater—*Touch the Stars* is a film depicting exploration by robotic spacecraft throughout the solar system, and *Space Explorers: The ISS Experience* is a four-part series shot aboard the ISS.

36. Take the NASA tram tour to see the historic mission control from the Apollo program. Space Center Houston.

Other tours require additional fees, including an expert-guided tour of the space center, breakfast with an astronaut, astronaut mission memories, and the Level 9 VIP tour offering guided behind-the-scenes access to Johnson Space Center. All the above activities were available at time of research, but check the center website for current offerings.

Countdown: Open daily; Monday–Friday 10:00 a.m.–5:00 p.m.; Saturday & Sunday 10:00 a.m.–6:00 p.m. (opens 9:00 a.m. some days, check website for updates)

Mission budget: $$$

Flight team: 281-244-2100 | https://spacecenter.org/

Coordinates: 29.5514, -95.0978 | 1601 NASA Parkway, Houston TX 77058

Flight plan: Space Center Houston is about 20 miles southeast of downtown Houston, just off I-45.

Rations: Space Center Houston offers several food options, including **The Food Lab Market & Eatery**, plus a coffee shop, food truck, and BBQ kiosk.

Orbital neighbors:

About two hours northeast of Houston, in the town of Lufkin TX, the small **Naranjo Museum of Natural History** has one of the largest Moon rocks on display anywhere, at 7.6 pounds, sealed inside a Lucite pyramid.

In far western Texas Hill Country, the University of Texas **McDonald Observatory** is a mountaintop astronomy complex open for public visitation and viewing. Daytime activities include a visitor center with exhibits, solar viewing, and self-guided walking tours to two summit-top observation domes. Nighttime activities include constellation tours and telescope viewing star parties. Events frequently sell out, and tickets must be purchased online in advance. **Flight team:** https://mcdonaldobservatory.org.

Frontiers of Flight Museum, Dallas TX

Debriefing: Located in a large hangar at Love Field, this facility is home to thirty aircraft, thousands of artifacts, and a small exhibition of spacecraft and missiles that tell the story of flight. Over twenty galleries and displays focus primarily on aviation plus several worthy space exhibits and artifacts. Aviation highlights include a full-size replica of the 1903 *Wright Flyer*, a Chance-Vought V-173 "Flying Pancake," a T-38A Talon trainer, and the ability to board and closely inspect a Southwest Airlines 737. A small missile gallery focuses mostly on rocket-powered air-to-air and cruise missiles.

The STARS: Space highlights include the authentic *Apollo 7* command module, which flew on a 1968 shakedown mission in Earth orbit—for crewed testing—only a few months before the ambitious *Apollo 8* mission. Other Apollo artifacts include two spacesuits from *Apollo 7*, a Moon rock from *Apollo 15*, and a lunar module hammock.

Additional artifacts include two cameras from Project Gemini including a Hasselblad and an instrument observer camera from *Gemini* 2, plus items from *Skylab* and the space shuttle program.

Nearby there's a pair of high-quality, full-size replicas including Sputnik, humanity's first satellite, launched by the Soviets in 1957, and *SpaceShipOne*, the first privately developed spacecraft to reach suborbital space in 2004.

A series of scale models, related to many NASA programs, depicts X-planes, spacecraft, and rockets. Elsewhere, there's a one-tenth model of the BA-2C, a never-flown rocket developed in the late 1990s by private company Beal Aerospace. Before shutting down in 2000, the company successfully test fired the most powerful liquid-fueled rocket engine since the F-1 from the Saturn V.

Countdown: Monday–Saturday 10:00 a.m.–5:00 p.m.; Sunday 1:00–5:00 p.m.

Mission budget: $

Flight team: 214-350-3600 | https://www.flightmuseum.com

Coordinates: 32.8427, -96.8353 | 6911 Lemmon Avenue, Dallas TX 75209

Flight plan: The museum is at **Dallas Love Field Airport**, several miles northwest of downtown Dallas.

Rations: Restaurants can be found throughout the area.

SpaceX Starbase Launch Viewing, South Padre Island TX

Debriefing: The SpaceX Starbase is a private spaceport, rocket production facility, and testing center located at Boca Chica—on the Gulf Coast, in southwestern Texas—near the border with Mexico. Starbase is intended to launch and recover the fully reusable *Starship*. If completed as planned, this two-stage vehicle, which burns methane and liquid oxygen, will become the tallest and most powerful rocket ever built. Incremental goals include launching Earth orbital missions, returning to the Moon, and ultimately exploring and colonizing Mars.

SpaceX does not allow public tours or launch viewing at the Starbase. But the southern end of South Padre Island is only five miles away from the launchpads, and there are several viewing locations on the island. The recommended site is the Isla Blanca Park, an RV and day-use facility, and this adventure is based on that option.

The STARS: The recommended viewing location is the closest public space to the Starbase launchpads. Isla Blanca County Park is a large beachfront camping and day-use facility with 600 RV sites with hook-ups, cabana rentals, and tent sites. The park has restroom and shower facilities, shaded picnic

37. The SpaceX Starship and Super Heavy booster at Starbase in Texas. Official SpaceX Photos.

123

areas, a water park, a full-service marina, and several restaurants. Many parts of the park offer unimpeded views of the Starbase launchpads, about five miles away.

During launch windows, the island gets busy, and reservations are available for short-term or long-term stays. Day use is available for a per-vehicle fee, but spaces are limited, so arrive early. Because launches may be postponed or rescheduled on short notice, most aspiring launch viewers spend several days, around the anticipated launch date, on the island. Determining precisely when a Starbase launch will happen is challenging, often involving piecing together information from multiple sources, which becomes clearer as the date approaches.

One source of upcoming launch info is www.rocketlaunch.live/?filter= spacex

For general information about the island visit www.sopadre.com/spacex.

For those who are curious but can't visit, there are Starbase live webcams: www.youtube.com/channel/UCFwMITSkc1Fms6PoJoh1OUQ

Countdown: Open daily

Mission budget: $$-$$$$

Flight team: 956-761-5494 | www.cameroncountytx.gov/parks-coastal -parks/parks-isla-blanca/

Coordinates: 26.0755, -97.1630 | 33174 State Park Road 100, South Padre Island TX, 78597

Rations: There are restaurants at the park and many more options around the island.

13 Bold Steps and Tragic Setbacks

The Start of Apollo

With the arrival of 1967 came the approach of Apollo's first crewed launch. Since Kennedy's announcement in 1961, the program had spent six years designing and testing the rockets, spacecraft, and equipment needed to land on the Moon.

Von Braun had launched ten test flights of the first-generation Saturn I rocket, often with uncrewed boilerplate command modules. He'd conducted three test launches of the more powerful Saturn IB, which used an uprated first stage and the new S-IVB second stage, which burned liquid hydrogen and liquid oxygen. Together, the IB more than doubled the low Earth orbit payload capacity, to 41,000 pounds.

Using this increased capacity, twice the IB had carried the complete cylindrical service module. This included the conical command module, with flight controls and room for three astronauts. The combined Command and Service Module (CSM) carried consumables, power supplies, thrusters, and a single AJ10 engine, which used hypergolic propellants, hydrazine and nitrogen tetroxide. One flight had tested the S-IVB upper stage in orbit, to verify design parameters for restarting its powerful J-2 engine in the vacuum of space.

An All-Up Approach

The time had arrived for two major milestones: a manned test flight in low Earth orbit, followed by the first test launch of the behemoth Saturn V rocket that would eventually carry the entire spacecraft and three astronauts toward a lunar landing. But to speed development, and meet Kennedy's challenge, NASA had adopted a somewhat radical policy. In 1963 James Webb had hired George Mueller, a missile-program manager, to become an associate administrator for NASA. Mueller took over coordination of three key facilities, the Manned Spaceflight Center in Houston, the Kennedy Space Center in Florida,

38. Gus Grissom, Ed White, and Roger Chaffee of the *Apollo 1* crew. NASA.

and von Braun's Marshall Spaceflight Center in Huntsville AL. Mueller's controversial decision was introducing the concept of all-up testing.

Until this time, von Braun had followed an incremental testing model, first used with the V-2 in Nazi Germany and later with the Redstone for the U.S. Army and with Project Mercury for NASA. They started by launching a first stage topped by dummy upper stages, sometimes empty or carrying ballast. As each stage was perfected, they moved on to another test launch by adding the next live stage. But Mueller recognized that if Apollo was going to land on the Moon within the decade, a faster approach was needed.

Another benefit to the all-up approach was that it saved money. This was an important consideration given the ballooning costs of Apollo. All-up testing was the method used by Korolev and the Soviets, who had limited resources and were conveniently able to conceal the inevitable all-up failures. All-up testing had successfully been used by the air force to build the Titan II, which launched the *Gemini* missions, and the Minuteman missile.

But those rockets were a fraction of the size of the Saturn V. When all-up testing failed, it was often a catastrophic explosion of the entire vehicle. Mueller had shrugged. So what? Von Braun considered all-up testing too reckless, but after some resistance, he acknowledged the benefits of this method. Under von Braun's incremental approach, it might have taken more than ten test flights before the Saturn V was qualified for human use. With Mueller's all-up testing method, it could be done with just a few launches. On the mission calendar, 1967 was penciled in as the year to find out if all-up testing would work.

Apollo 1

The first crewed flight was labeled AS-204, with a Saturn IB launching an Apollo CSM into low Earth orbit. Gus Grissom, a veteran of Mercury and Gemini, would command. He was joined by America's first spacewalker, Ed White, and newcomer Roger Chaffee.

On January 27, less than a month before launch, the three-man crew was sealed inside the command module for a "plugs out" test, when the spacecraft would operate under its own power for the first time. The three-piece hatch was installed, held in place by interior pressure from a pure oxygen environment running a few PSI higher than outside. During the simulated countdown, a series of problems developed, including sporadic communications.

"How are we going to get to the Moon," joked Grissom, "if we can't talk between two or three buildings?"

Ground control didn't have an answer, and the channel was filled with static for a minute. Suddenly the radio erupted in a series of shouts.

"Flames!"

"We're fighting a fire in the cockpit."

"We have a bad fire. Open 'er up!"

The final sound over the radio was a scream of horrific pain, as smoke and flames engulfed the three astronauts. In their last moments, the men clawed at the hatch, but it opened inward and was sealed shut by the rising pressure. A few seconds later, the command module ruptured, and smoke filled the service structure. It took five minutes for technicians to open the hatch, finding all three astronauts dead inside from smoke inhalation and severe burns.

At the request of the three men's widows, the flight was renamed *Apollo 1* to honor the fallen astronauts. Crewed flights were suspended, while NASA convened an investigation. In April, the review board released their report. The likely ignition source was poorly insulated wiring inside the capsule. From there, the pure oxygen atmosphere helped fuel the flames, along with highly combustible materials found throughout the cabin. Even the astronauts' nylon spacesuits had melted, fusing their bodies to interior surfaces. The flawed hatch design had prevented them from escaping.

Hundreds of improvements were needed before the crewed spacecraft could fly, including an interior overhaul and an outward opening hatch. Cabin materials and spacesuits needed to be redesigned using nonflammable materials. And NASA was criticized from outside and within for being too gung-ho and ignoring problems that had arisen. Despite these concerns, there was little talk about cancelation and Apollo continued.

Soviet Setbacks

Meanwhile, the Soviet program was experiencing tragedies of their own. In early January 1966, Korolev had gone to the hospital for an uncertain reason. Since his time in the Siberian gulag, he'd developed a number of ailments, including issues with his heart and other organs. Nine days later, the chief designer died. Only afterward was his identity revealed to the world. Sergei Korolev's ashes were interred in the Kremlin Wall with other national heroes of the Soviet Union.

The Soviet space program would have to forge ahead without their accomplished leader. The first step was getting the Soyuz spacecraft into orbit. After a series of unmanned test flights, the first manned launch was scheduled for April 23, 1967. But soon after *Soyuz 1* was launched, with Vladimir Komarov aboard, problems began to mount. The solar panels didn't fully deploy, causing a shortage of power. Issues developed with improper spacecraft orientation. After eighteen orbits, the retro rockets were fired. But during descent the parachutes malfunctioned. The capsule slammed into the ground in southwestern Russia, and Komarov was killed on impact.

Like the Apollo program, which had been similarly suspended, the Soviets grounded further Soyuz flights while conducting a comprehensive safety overhaul. Now the question was which program might resume crewed flights first in their quests to reach the Moon.

39. Plumes of debris rise from explosions at the Cinder Lakes crater field near Flagstaff AZ. USGS.

When NASA Bombed Flagstaff

In late July of 1967, the high desert in northern Arizona was rocked by a series of explosions. Debris flew hundreds of feet into the air, with plumes of cinders and dirt crisscrossing beneath the San Francisco Peaks. To those who didn't know the story, it would have appeared like the outskirts of Flagstaff were being bombed.

In recent years, NASA had been searching for an astronaut training ground on Earth that simulated the topography of the Moon. The unmanned lunar orbiter program was nearing completion of a detailed surface map. From early results, five potential landing sites were selected, with the leading candidate located in the Sea of Tranquility.

To create a lunar analog, the USGS Astrogeology Science Center in Flagstaff suggested modifying Cinder Lake, a field of volcanic cinders that erupted from Sunset Crater a thousand years ago. The loose volcanic material, mostly ranging in size from sand to pebbles, was thought to resemble the lunar surface astronauts might encounter during a landing.

129

To make the Cinder Lake crater field match its lunar counterpart, holes were dug to precise depths and a calculated amount of explosives was buried in each hole. The first bombardment was the biggest, followed by two more of decreasing size. When the hundreds of detonations were over, the expanse of black volcanic cinders had been transformed into a field of blast craters that closely matched landing site II P-6-1.

Over the coming years, scientists, engineers, and training astronauts descended upon the field. A mock-up of the lunar ascent module was erected at the proper height. Wearing mock-up spacesuits, workers developed and practiced lunar soil and rock sampling procedures. Lunar rover excursions were simulated using a test vehicle called Grover.

Elsewhere in the region, the training continued. At nearby Lowell Observatory, astronauts peered through telescopes at the lunar surface and studied Moon maps. Spacesuits were tested amid the sharp-edged rocks at Bonita Lava Field—when the suits ripped, they were redesigned.

At Meteor Crater near Winslow AZ, the astronauts studied one of Earth's best-preserved impact sites. Around 50,000 years ago, a 160-foot metallic meteorite struck present-day Arizona at an estimated 40,000 mph, excavating a crater almost three-quarters of a mile wide and 600 feet deep. With scientists leading the way, astronauts scrambled around the crater floor, slopes, and rim. This offered a chance to study ejecta patterns and develop an understanding of how to collect crater samples.

Other astronauts and scientists, including the *Apollo 11* crew, hiked to the bottom of the Grand Canyon on the South Kaibab Trail. Along the way, they collected rocks and studied sedimentary formations to learn about geology. After a night at Phantom Ranch, they hiked back up to the South Rim on the Bright Angel Trail. At the atomic bomb craters of the Nevada Test Site, the astronauts learned about secondary craters, which are formed by the impacts of ejecta from much larger primary craters. And to learn to identify volcanic structures, astronauts explored lava vents, lava tubes, and lava lakes in Hawaii.

The bulk of training, however, was far more technical: hundreds of hours in spacecraft simulators communicating by radio, flipping switches, and troubleshooting problems. Wearing mock-up EVA (extravehicular activity) suits, the astronauts simulated lunar excursions in underwater tanks. Elsewhere, they conducted g-force training, photography training, and helicopter training. Perhaps most harrowing was a type of lunar landing simulator nick-named the Flying Bedstead. The Lunar Landing Training Vehicle (LLTV) was a vertical-take-off-and-landing aircraft with a single jet engine on a gimbal pointed downward to simulate the flight characteristics of the lunar module. Using hydrogen peroxide thrusters to maneuver, the LLTV was critical for preparing NASA astronauts to land on the Moon. But it was also unwieldy and dangerous. In 1968 Neil Armstrong narrowly avoided disaster by eject-ing from a first-generation research version and parachuting to a hard but safe landing nearby.

40. A Saturn V rises as seen from the launch tower. NASA.

14 Closing In

The Saturn V Roars

While modifications to the Block II command module continued throughout 1967, the next uncrewed test flight went ahead. In early November, the Saturn V rocket rolled out on a crawler transporter from the massive Vehicle Assembly Building. At the launchpad, it took almost 150 truckloads and rail cars to fill the fuel and oxidizer tanks. When ready, the 36-story rocket weighed about 6.5 million pounds, with around 95 percent being propellants.

It was called *Apollo 4*, and lift-off was planned for 7:00 a.m. on November 9. This would be NASA's first all-up launch, and the first flight for the S-IC first stage and S-II second stage. Inside Kennedy Space Center, von Braun and the Apollo team watched. Outside, the bleachers were filled with NASA personnel, government officials, diplomats, and reporters awaiting the launch of the most powerful rocket ever built. At T-minus eight seconds, the five F-1 engines ignited with a thunderous roar, throttling up to 7.6 million pounds of thrust. Inside the buildings, the vibration was so intense that dust and ceiling tiles fell on observers.

The hold-down arms released, and twelve seconds later, the rocket cleared the tower. Burning over 3,500 gallons of propellant every second, acceleration increased as the weight of the rocket decreased. About a minute into the flight, and four miles up, the rocket broke through the sound barrier at around 770 mph. As the S-IC first stage neared depletion of its fuel, the Saturn V was now at an altitude of just over forty miles, traveling faster than 6,100 mph. The center engine cut off, and fifteen seconds later the remaining four outboard engines shut down. Small explosive charges separated the first and second stages. Small solid-fueled retro rockets fired, located in the vehicle's fairing assembly near the tailfins. This reverse thrust backed the S-IC first stage away, which then continued on a ballistic path toward the sea.

Now it was the S-II second stage's turn. First, small solid-fueled ullage

41. Earth as seen during the *Apollo 7* mission. NASA.

rockets, mounted on the aft interstage skirt, fired, increasing separation from the detached first stage. The other purpose of this extra momentum was pushing fuel into the lines to prepare for ignition. Next, the interstage skirt, a steel cylinder connecting the stages, separated with small explosive charges, revealing five J-1 engines. Generating over 1.1 million pounds of thrust, the second stage fired for about six minutes, reaching a speed of 15,600 mph and an altitude of around 110 miles.

Now, the S-II second stage separated from the S-IVB third stage, following a similar method to the first-stage separation. As the S-II ballistically arced toward the sea, the S-IVB fired its single J-1 engine for about 2.5 minutes, reaching an orbital velocity of nearly 17,500 mph at about 118 miles up. This first burn used about a quarter of the S-IVB's propellants, which left plenty for the second burn.

After two orbits, the S-IVB executed the world's first in-space reignition of a rocket. In a simulation of a translunar injection burn, the J-1 engine initiated a burn that put the spacecraft into a highly elliptical orbit, with an apogee of over 1000 miles and a perigee, or low point, purposely set 50 miles below the earth's surface. Shortly after the second S-IVB burn, the CSM separated and fired the service module's engine. The goal was to increase the speed to just under 25,000 mph, simulating an inbound return from the Moon. The command module separated and plunged through the atmosphere, heat shield first. The parachutes deployed and it landed successfully in the Pacific Ocean near Midway Island. It was a flawless test flight. The Saturn V had worked.

A Turbulent Year

As the crewed flight hiatus continued, two more unmanned test launches went forward in early 1968. After many delays, the first complete lunar module was finally ready to go in January. Fitted atop a Saturn IB, the *Apollo 5* launch and separation went well, putting the lunar module into a standard orbit. Despite some mishaps with fuel supply and the guidance computer, the ground team was able to control the craft remotely. They fired the descent engine, separated the lunar module, and fired the ascent-stage engine. After they completed the mission objectives, the various stages were left to burn up in the atmosphere upon reentry.

A few months later, *Apollo 6* was the second test launch of the Saturn V. A major goal was to put the S-IVB and CSM on a translunar injection toward the Moon. Shortly after separation, the CSM would turn around and fire the service module engine for a direct-return abort back to Earth. But during launch the S-IC first stage experienced severe pogo oscillation due to thrust fluctuations. Like a pogo stick, the rocket essentially accelerated and decelerated along its vertical axis by +/- 0.6 g, more than twice its designed tolerance.

After first-stage separation, these problems continued with the S-II second stage, with two of the J-2 engines shutting down prematurely. The guidance computer in the instrument unit, an IBM-built ring fitted atop the S-IVB third stage, compensated by burning the remaining J-2 engines for an extra minute. After the S-II jettisoned, the S-IVB burned for an extra thirty seconds to reach a less than optimal elliptical orbit, just high enough to continue the mission. But after two orbits, the S-IVB failed to reignite for its second burn for translunar injection. Mission control shifted to a repeat of the *Apollo 4* test, successfully firing the service module engine for a high-speed command module reentry.

Post mission, the J-2 failures were traced to ruptured fuel lines, which would need to be strengthened. And to solve the pogo problem, von Braun's Saturn V team went to work on dampening vibration within the propellant tanks using baffles. As work continued, the improvements were deemed successful enough to qualify the Saturn V for human flight.

Unlike the first flight of the Saturn V, little attention was paid to *Apollo 6* by the press. On the very same evening as the launch, April 4, 1968, Martin Luther King Jr. stepped out on a balcony of a Memphis motel and was killed by an assassin's bullet. Recognized as the leader of the civil rights movement, King had advocated for nonviolent protest against racial inequality. In the aftermath, many cities erupted into fiery riots. Meanwhile, the war in Vietnam was spiraling into chaos. And just two months later, while campaigning for president, JFK's brother Robert Kennedy was assassinated after an event in California. The year would go down as one of the most turbulent in American history.

The Zond Program

Throughout 1968 the Soviet Moon program was working feverishly to upstage Apollo. Korolev's super heavy-lift N-1 rocket, offering the massive thrust needed for a lunar landing mission, was still lagging in development. And while the chief designer's legacy rocket, the R-7, had been upgraded to launch the heavier Soyuz spacecraft, it didn't have the power needed to send a crewed mission to the Moon.

With regards to the Soyuz, the Soviets were ready to move past the tragic death of Komarov in the crash of *Soyuz 1* the year before. In October, an upgraded *Soyuz 2* launched as an unmanned rendezvous target for *Soyuz 3*, which made it to orbit one day later with a single cosmonaut. Following this successful test flight, preparations were made for more ambitious flights of the Soyuz. The Soviets now had a versatile spacecraft that could carry cosmonauts to the Moon; they just needed a rocket to get them there.

The death of the chief designer several years before had opened up opportunities for new solutions from other Soviet design bureaus. In particular,

Vladimir Chelomei was a missile designer and leader of OKB-52, located east of Moscow. Chelomei had recently designed the large UR-500 Proton to deliver a 100-megaton thermonuclear warhead. Standing 174 feet tall and weighing around 1.5 million pounds, the rocket had a clustered first stage with six engines providing about 2.5 million pounds of thrust.

Having roughly a third the power of a Saturn V, the Proton was oversized for an ICBM, but it had plenty of potential as a space rocket. When he was in charge, Korolev had opposed rockets like the Proton as crewed launch vehicles for using hypergolic propellants, hydrazine and nitric oxide, which are highly toxic to humans. But some rocket designers felt differently, and hypergolic propellants had been used successfully in other crewed launches, including NASA's Project Gemini.

Now that Korolev was deceased, the winning argument was that the Proton had the potential to make a lunar voyage. With some slight improvements, the Proton-K rocket would have just enough power to send a stripped-down Soyuz capsule with two cosmonauts on a free-return trajectory around the Moon. The spacecraft would be called *Zond*, meaning *probe*. Launching without the forward orbital module, the two cosmonauts would be confined to the small descent module for the entire flight. They wouldn't land. They wouldn't orbit. They would just swing around and use the Moon's gravity to slingshot them back to Earth. But most importantly, they would do it first.

Uncrewed test flights began in 1967, and early results were mixed. The first *Zond* made it to orbit as planned. So did the second flight, but the Blok D upper stage failed to put the spacecraft on a translunar trajectory. The next two Proton-Ks veered off course during launch and crashed, though the escape towers carried the *Zond* capsules to safety.

In March 1968 *Zond 4* was launched on a simulated translunar trajectory away from the Moon. Similar to *Apollo 4* and *6*, the *Zond* entered a highly elliptical orbit, reaching an apogee of 220,000 miles, approximately lunar distance. But when the returning spacecraft hit the atmosphere, it failed to execute an atmospheric skipping maneuver to reduce speed. As the descent module ballistically plummeted toward the Gulf of Guinea, off West Africa, the capsule self-destructed to prevent its recovery outside Soviet territory. Two more launch failures followed, which meant more uncrewed tests were needed before risking the cosmonauts' lives. But despite these mishaps, the Soviets were closing in on the American program.

42. While orbiting the Moon on *Apollo 8*, NASA astronaut William Anders took one of the most famous photos in human history: Earthrise. NASA.

15 Around the Moon

Soviets Circling

The Zond program's first complete success came in September 1968. A Proton-K rocket launched successfully and Blok D reignited for a translunar injection. *Zond 5* coasted around the Moon at a distance of 1,200 miles. On board were the first living passengers to circle the Moon, including fruit fly eggs, plant cells, and two Russian tortoises. The capsule carried something else: a radio relay, which transmitted the voices of cosmonauts discussing flight data and their approach to the lunar surface. The transmissions were picked up by Earth observers, including the CIA, which feared the Soviets were about to land.

Three days later, the descent module did land—but on Earth. The atmospheric skipping maneuver had failed again. But a ballistic descent allowed a backup landing in the Indian Ocean, where Soviet naval forces had been waiting. The turtles were safely recovered, though the high speed of reentry, with deceleration forces estimated up to 20g, was likely too high for human survival.

The Soviets had scored yet another Space Race victory with this first lunar circumnavigation and safe return to Earth. But another uncrewed mission was needed before risking two cosmonauts' lives. In early November, *Zond 6* followed with a repeat of the previous mission. Another successful circumnavigation also included a radio broadcast of Russian voices discussing a lunar landing.

Upon return, this time the atmospheric skipping worked, and the descent module came down over Soviet territory. However, module depressurization had already killed the animals onboard, and a parachute malfunction at several miles high meant the capsule crashed into the ground. The Zond program was close, but more work was needed. To international observers, it appeared likely the Soviets would attempt a crewed mission within months.

Apollo Resumes

After a twenty-one-month hiatus due to the *Apollo 1* fire, crewed missions resumed in late 1968. The fate of the Apollo program, and the ability to meet Kennedy's challenge, seemingly rested with the success of this long-delayed mission. In response, NASA turned to experienced Mercury and Gemini astronaut Wally Schirra to command.

On October 11 the three-man crew was sealed inside a redesigned command module and launched atop a Saturn IB rocket. The eleven-day mission was packed with orbital tests of the spacecraft, which had undergone over a thousand changes. The docket included navigation techniques, rendezvous using the service module's SPS engine, and verifying the habitability of the spacecraft. From a technical standpoint, the mission was flawless.

But even before launch, tensions were brewing between the flight crew and mission control. During countdown, high-altitude winds came in from the east. In the event of an abort, the winds could carry the command module away from the sea for a hard inland touchdown that might cause injury. Despite Schirra's protests, Apollo managers disregarded the safety protocol and authorized the launch, which fortunately went perfectly.

During the first day in orbit, Schirra developed a head cold, which soon spread to his crewmates. With headaches from sinus congestion, made worse by microgravity, the irritable crew became increasingly resentful of a demanding ground team adding tasks to their already busy schedule. When mission control asked for an unplanned test of the onboard TV system, Schirra refused and stated all broadcasts would be postponed, to focus on more critical mission objectives. The ground team was stunned but yielded.

Later in the mission, the rescheduled live TV broadcast, the first from an American spacecraft, went well. The amiable crew shrugged off their discomfort and gave a smiling tour of the spacecraft, offering NASA some much-needed positive publicity. But behind the scenes, the tension continued. The crew wasn't sleeping well, and though they carried out every test asked of them, Schirra complained that some were idiotic. When it came time to land, they refused to wear their helmets, worried the changing pressure would rupture their eardrums without the ability to pinch their nostrils to equalize. After landing without helmets, the soon-to-retire Schirra was hauled into the astronaut office for a lecture about what Deke Slayton later called the first "space war."

A Wild Idea: Apollo 8

During the summer of 1968, Apollo manager George Low had suggested a wild idea for the *Apollo 8* mission, originally intended to test the lunar module in Earth orbit. As the flight approached, the lunar module simply wasn't ready to fly. Instead of repeating the objectives of *Apollo 7*, why not send the *Apollo 8* lunar module on a crewed flyby to the Moon?

Such a mission would keep the Apollo program moving forward, and hopefully beat the Soviet Zond program to the first crewed lunar circumnavigation. One by one, the other program managers agreed, including von Braun. It was a bold plan, involving a great deal of last-minute preparation, but one they felt was within the program's capabilities. The originally scheduled crew for *Apollo 9*—Frank Borman, Jim Lovell, and William Anders—were moved up a flight and began training around the clock for translunar navigation.

On December 21, the first crewed flight of the Saturn V rocket successfully launched from Cape Kennedy. From a nearly circular low Earth orbit, the S-IVB third stage perfectly reignited for a translunar injection burn of 318 seconds. Soon the CSM was hurtling toward the Moon at just over 24,500 mph—the fastest speed humans had ever traveled. Once en route, the CSM separated from the S-IVB. Using thrusters, the astronauts turned their spacecraft around to take photos. In the distance, the entire globe of the earth was visible to people for the first time in history.

During the sixty-eight-hour outbound flight to the Moon, the crew was mostly unable to see their destination due to fogging of the windows and the orientation of the spacecraft. They monitored their course using star sightings, made a few television broadcasts, and tried but mostly failed to sleep. Borman briefly became sick, at the time attributed to a stomach bug but, in hindsight, NASA's first case of space sickness.

To enter lunar orbit, the crew would have to perform a retrograde burn with the service module engine to slow the spacecraft down to about 3,600 mph. Due to orbital mechanics, this event had to occur on the far side of the Moon, out of radio contact with Earth. On the mission's third day, they sailed around the far side. The four-minute and seven second burn began in the Moon's dark shadow. Nailing the precise duration was critical. If the engine stopped too soon, they might fling past the Moon into space. If the engine burned too long, they could lose too much speed and impact the Moon. Fortunately, the engine stopped right on time. *Apollo 8* passed into the sunlit side and finally the crew caught their first glimpses of the lunar surface, below.

Another World

"The Moon is essentially gray, no color," reported Lovell, observing through a window. "Looks like plaster of Paris or sort of a grayish beach sand. We can see quite a bit of detail." Lovell spent the next few minutes describing the many craters, including newer ones he believed had been hit by meteorites or projectiles. One huge crater, named Langrenus, had terraced walls. Another crater, called Pickering, had faint rays of ejecta radiating outward along the surface. As the Sea of Tranquility approached, Lovell focused on spotting landmarks associated with the proposed first landing site, including a triangular mountain he would later name for his wife, Mount Marilyn.

Apollo 8's initial orbit around the Moon was elliptical, roughly 200 by 70 miles. So, on the second orbit, they conducted another retrograde burn of the SPS for a brief eleven seconds to slightly slow the spacecraft. This nearly circularized their roughly two-hour orbit at about 70 miles altitude. During the fourth orbit, the crew came out from the far side and witnessed Earth rising above the lunar horizon. Anders snapped a color photo that would become one of the most famous images ever captured: Earthrise.

As *Apollo 8* continued to circle the Moon, the crew was exhausted. They'd slept little during the first few days, and now Borman ordered them to take turns napping. When the spacecraft circled the Moon for the ninth time, the crew began their second TV broadcast. It was Christmas Eve. Back on Earth, it is estimated that up to a quarter of all humanity eventually saw the footage. Roughly a billion people, many of them gathered around televisions.

Taken with the camera pointed out the rendezvous window, the grainy black-and-white image showed a portion of the curved Moon rotating below. After Borman introduced the crew, each astronaut offered his impressions of the lunar surface.

"A vast, lonely forbidding type existence," said Borman. "Or expanse of nothing, that looks rather like clouds and clouds of pumice stone."

"The vast loneliness up here, of the Moon, is awe inspiring," said Lovell. "And it makes you realize just what you have back there on Earth. The Earth from here is a grand oasis in the big vastness of space."

"The thing that impressed me the most was the lunar sunrises and sunsets," said Anders. "These, in particular, bring out the stark nature of the terrain, and the long shadows really bring out the relief that is here."

As light and dark patches of the lunar surface passed below, the astronauts proceeded to describe the features they saw. Passing beneath them were many mysteries the Apollo program hoped to solve. Double-ring craters that looked like bullseyes. Dark areas called maria, theorized to be old lava flows. Delta-rim craters that lacked ejecta blankets and appeared to be filled with the same material as lunar maria. Perhaps the most enigmatic formations were the lunar rilles. These sinuous channels looked like dry riverbeds, winding across the surface.

As the spacecraft approached the terminator, the border between dark and light that delineated the lunar sunrise, the astronauts concluded their broadcast by reading a few verses from the Bible's book of Genesis. "And from the crew of *Apollo 8*," said Borman, "we close with good night, good luck, a Merry Christmas—and God bless all of you, all of you on the good earth."

After ten orbits over twenty hours, *Apollo 8* fired its engine for a trans-Earth injection. Two and a half days later, they began the reentry. Upon a successful splashdown in the Pacific, the astronauts were greeted as world heroes. They were hailed for providing an uplifting ending to a tumultuous 1968 in America, and *Time* magazine named them the Men of the Year.

The response from Brezhnev and the U.S.S.R. was to congratulate the Americans for their historic achievement. Publicly, they denied ever considering sending their own cosmonauts. For them, the official line became that lunar missions were a political stunt. After dominating the Space Race for over a decade, the Soviets were internally reckoning with their first major loss. It seemed likely the Americans could land astronauts on the Moon within months. Though perhaps there was still a way for the Soviets to spoil this seemingly inevitable landing.

Sites in the Midwest
Nine Adventures and Seven Orbital Neighbors

Celestial Overview: The Midwest is a vast region spanning a dozen states, give or take. But luckily several of the space highlights are concentrated in a few areas. A pair of must-see sites can be found in Ohio. One is the National Museum of the U.S. Air Force, located in Dayton. This massive facility has an impressive exhibition of aircraft, spacecraft, rockets, and missiles. About a three-hour drive away, in Cleveland—a short jaunt in Midwestern terms—there's another must-see space museum. The Great Lakes Science Center is also the NASA Glenn Visitor Center, combining authentic vehicles and artifacts with interactive exhibits about science and space, and just down the street is the International Women's Air and Space Museum. Rounding out the state's offerings are several smaller but worthy museums and two astronomy sites. Plus a pair of excellent space sites can be found just over the state line in Pittsburgh, discussed in the Northeast region chapter.

A third must-see destination is in Chicago. The Museum of Science and Industry is a mammoth facility with many remarkable artifacts and exhibits, including an authentic German U-boat. Meanwhile, the space gallery offers space-flown vehicles, artifacts, and displays. Within walking distance is the Adler Planetarium. Rounding out the Midwest region, further space adventures can be found in St. Louis, southern Michigan, and rural Indiana.

Great Lakes Science Center and NASA Glenn
Visitor Center, Cleveland OH

Debriefing: Located on the waterfront of Lake Erie, in downtown Cleveland, the Great Lakes Science Center is an interactive museum with exhibits spanning biology, technology, and space exploration. In 2010 the NASA Glenn Visitor Center relocated to the science center from its former location at NASA Glenn Research Center, a development facility for space exploration and propulsion technologies. In addition to many space highlights, the center offers exhibits about wind and solar power, biomedical technology, and the science of rock and roll music. Visitors can also tour the William G. Mather, a restored 618-foot Great Lakes steam freighter from the 1920s that's docked beside the museum. There's a dome theater that shows rotating films, some related to space topics.

The **STARS:** Three galleries focus on space topics. The Explore gallery presents the history of NASA and the space age. A highlight is the space-flown

43. At the Great Lakes Science Center in Cleveland you can see the airbag system used to land the *Pathfinder* spacecraft and *Sojourner* rover on Mars. Joe Yachanin/Great Lakes Science Center.

capsule from the *Skylab 3* mission in 1973, one of the few Apollo command modules left unenclosed, which allows for closer inspection.

Other artifacts include a second-gen J-mission Apollo spacesuit, a Moon rock from *Apollo 15*, a lunar rover mesh tire test article, and Buzz Aldrin's flight suit from *Apollo 11*. A display and several items relate to Ohio native John Glenn's missions, including the *Freedom 7* Mercury flight and his return to space on the shuttle in 1998. Another display focuses on NASA shuttle astronaut Guion Bluford, who became the first African American in space in 1983. Dispersed throughout the galleries are scale models ranging across NASA programs, including recent vehicles like the Ares I-X and *Orion* spacecraft.

The Living in Space gallery offers exhibits and artifacts about day-to-day life in microgravity, including eating, sleeping, and using the bathroom. Other displays focus on health and exercise in space. A scale model of the International Space Station (ISS) is accompanied by a live feed of Earth as seen from space and a map showing the current orbital location of the ISS. Suspended from the ceiling, there's an airbag system developed at the research center for landing the *Pathfinder* mission on Mars, including the first rover, *Sojourner*.

The Discover gallery combines hands-on exhibits with displays about the contributions of the Glenn Research Center to NASA space endeavors. Exhibits relate to wind tunnel research, corresponding test articles, space shuttle aerodynamics, and more. A presentation on solar power and panels includes a partial array from the ISS. Other highlights include an exhibit on deep space communications and a cutaway jet engine and rocket motor allowing visitors to examine interior components.

Throughout the galleries, exhibits and demonstrations explore many topics, such as preparing meals in space and developing materials, including micrometeorite shielding and thermal tiles like those used on space shuttles and other spacecraft. A pair of unique labs simulate lunar and Martian soils, while others relate to aerodynamics and rocketry.

EVAs: The Discovery gallery includes two flight simulators, one for landing on the Moon and the other for landing a space shuttle.

Countdown: Tuesday–Saturday 10:00 a.m.–5:00 p.m.; Sunday 12:00–5:00 p.m.; closed Monday

Mission budget: $

Flight team: 216-694-2000 | https://greatscience.com

Coordinates: 41.5073, -81.6966 | 601 Erieside Avenue, Cleveland OH 44114

Flight plan: Located just off the Memorial Shoreway in downtown Cleveland, and paid parking is available at the center's attached facility.

Mission parameters: Just across the plaza is the **Rock & Roll Hall of Fame**, while if you keep walking east, you'll reach the free **International Women's Air and Space Museum**.

Rations: With views of the harbor, **Spark Kitchen** is the center's on-site restaurant. Further options can be found throughout downtown.

International Women's Air and Space Museum, Cleveland OH

Debriefing: Located at Burke Lakefront Airport, this museum explores the impact of women in aviation and space exploration. Many exhibits focus on individual aviators, including Amelia Earhart, Ruth Nichols, Bessie Coleman, Harriet Quimby, Katharine Wright, Jackie Cochran, the Women Airforce Service Pilots from World War II, and more. Other exhibits focus on women involved in space exploration.

The STARS: A variety of exhibits explore women in space travel; some are permanent while others rotate yearly. One highlight is an exhibit focusing on each of the women included in and recruited for the *Mercury 13*. They were female aviators who, in the 1960s, underwent a privately funded demonstration program for astronaut training based upon Project Mercury.

A comprehensive Space Wall features regularly updated displays about all women who have been to space, including the first three, Valentina Tereshkova, Svetlana Savitskaya, and Sally Ride, plus more recent commercial astronauts like Beth Moses of *SpaceShipTwo*.

Other exhibits discuss the 1978 NASA astronaut class, the first to admit women, and the women who have worked at mission control, including Joanne Morgan, the only woman in the room for the *Apollo 11* launch, plus her post-mission life.

Women's roles on the International Space Station are presented with topics like living and working in space. Other displays discuss human health in space, exercising in space, and conducting microgravity experiments. The museum is currently expanding their online virtual tours and YouTube offerings for those unable to visit.

EVAs: Guided tours are available for a small fee by appointment only. Call the museum at least two weeks in advance.

Countdown: exhibits open daily 8:00 a.m.–8:00 p.m.; gift shop 10:00 a.m.–4:00 p.m.

Mission budget: Free

Flight team: 216-623-1111 | https://iwasm.org

Coordinates: 41.5113, -81.69 | Burke Lakefront Airport, 1501 North Marginal Road, Suite 165, Cleveland OH 44114

Flight plan: Located at **Burke Lakefront Airport**, just off Memorial Shoreway in downtown Cleveland. Paid parking available.

Orbital neighbor: In Hocking Hills State Park, about an hour south of Columbus, the **John Glenn Astronomy Park** offers guided stargazing programs on Friday and Saturday evenings from March through November. On clear nights, the programs start about thirty minutes after sunset with guides using laser pointers to explore constellations and other night sky features. Topics of the shows vary throughout the year, from planets to familiar constellations to extragalactic space. The park is also popular during eclipses and meteor showers. Visitors are welcome to bring their own telescopes. Occasional day programs occur as well. Check the park website for an updated schedule. **Flight team:** https://jgap.info.

National Museum of the U.S. Air Force, Dayton OH

Debriefing: The National Museum of the U.S. Air Force is an impressive place, displaying over 360 aircraft, missiles, and spacecraft in a 1.1-million-square-foot facility. While most of the displays are naturally focused on aviation, there are many space highlights to be seen that make this an excellent destination for space enthusiasts. Aviation highlights include the restored Memphis Belle B-17 from World War II, a prototype XB-70 Valkyrie supersonic bomber, Air Force One from the 1960s and other presidential aircraft, an F-117 Nighthawk stealth fighter, and many more.

The STARS: Many highlights are found in the Lockheed Martin Space Gallery. There's a flight-rated Project Mercury capsule that never flew and instead provided parts for the final mission, *Faith 7* with pilot Gordo Cooper. There's an authentic Gemini B spacecraft from the USAF Manned Orbiting

44. A rocket and missile collection at the National Museum of the U.S. Air Force. USAF.

Laboratory program, a canceled orbital reconnaissance station. Another highlight is the space-flown command module Endeavor from *Apollo 15*, the only lunar mission with an entirely air force crew (David Scott, James Irwin, and Alfred Worden). There's also an exhibit of replica spacesuits from across NASA's manned spaceflight programs.

The space shuttle exhibit combines an actual crew compartment trainer (CCT) with a mock-up orbiter fuselage. Visitors can enter the CCT, which is one of three mock-ups that were used to train crews during the shuttle program. A number of satellites mostly includes U.S. reconnaissance orbital systems like the Gambit, Hexagon, DSP, and Teal Ruby.

Rockets include a horizontally elevated Titan IV-B, the air force's largest and most powerful launch vehicle. There's also an ASM-135A, a U.S. anti-satellite missile of the type that destroyed a failing solar observation satellite in a 1985 test. Displayed rocket engines include an XLR, like those used on the Bell X-1 and early X-15, and an XLR99, which powered upgraded X-15s on their record-breaking flights. There's also an LR79 liquid-fueled engine, like those from multiple launchers, including the Juno II and Thor, and an LR87 from the Titan ICBM series, like those used to launch the Project Gemini spacecraft.

45. An X-15, a Titan IV-B, and more at the National Museum of the U.S. Air Force. USAF/Ken LaRock.

Several authentic space planes include one of the two remaining NASA X-15s, from the 1950s and 1960s, that were flown to the edge of space, reaching a maximum altitude of 67 miles. (During 2022 the other remaining X-15 was temporarily off display at the Smithsonian Air and Space Museum, though it will likely be displayed again in the future.) There's also a Boeing X-40A, a robotic test vehicle from the USAF Space Maneuver Vehicle program that led to the operational X-37 space plane. Two other X-planes include the X-24A and X-24B, a pair of experimental rocket-and-glide lifting body aircraft from the 1970s that supported space shuttle development.

Another space-relevant aircraft is the C-119J Flying Boxcar, which made the world's first midair recovery of an object returned from space. In 1960 this specially adapted aircraft retrieved the first of many satellite film recovery capsules parachuting through the atmosphere.

Elsewhere are the gondolas from three high-altitude ballooning programs during the 1950s and 1960s: Project Man High, Project Excelsior, and Project Stargazer. Other exhibits focus on USAF space programs and women in space.

EVAs: Guided tours are free and offered daily, with topics varying throughout the day. Check the museum website for topics and times. The on-site theater offers various films throughout the day, plus there are several simulators and rides throughout the museum. Tickets for films and simulators can be purchased at the museum, while more information is available on the website.

Countdown: Open daily 9:00 a.m.–5:00 p.m.

Mission budget: Free

Flight team: 937-255-3286 | www.nationalmuseum.af.mil/

Coordinates: 39.7815, -84.1103 | 1100 Spaatz Street, Wright-Patterson AFB OH 45433

Flight plan: The museum is a few miles northeast of Dayton, near OH-4.

Rations: There are no restaurants at the museum, but options can be found throughout the city.

Orbital neighbor: The **Cincinnati Observatory** offers daytime history tours of the facility and nighttime telescope observations and shows on a variety of astronomy topics. **Flight team:** www.cincinnatiobservatory.org

Armstrong Air & Space Museum, Wapakoneta OH

Debriefing: Located in rural western Ohio, this museum tells the story of hometown hero Neil Armstrong and contributions to space travel by other Ohioans. Opened in 1972, the building is designed to resemble a futuristic Moon base. Exhibits have evolved over the years to include later programs like the space shuttle and uncrewed planetary probes.

The STARS: Outside the museum, there are full-size mock-ups of Apollo and Gemini capsules, the latter of which visitors can sit inside. There's also a restored F5D Lancer, which Neil Armstrong flew during test flights for the U.S. Air Force's brief X-20 Dyna-Soar space plane program.

Inside the museum, artifacts include the Gemini 8 capsule that Armstrong piloted during a harrowing 1966 malfunction in orbit that aborted the mission. Also on display are Neil Armstrong's Gemini and Apollo spacesuits, plus an *Apollo 11* Moon rock. Another displayed aircraft flown by Armstrong is the Aeronca Champion he learned to fly as a teenager.

Other exhibits include an H-1 engine like those used on the Saturn I and Saturn IB rockets, several space shuttle displays and artifacts, and a one-quarter-scale model of the Mars rover *Curiosity*. The model was built by engineering students at Ohio Northern University.

Included with admission, the museum's central dome houses the Astro Theater, which shows a twenty-five-minute documentary about the *Apollo 11* lunar landing. There are also two simulators, one for lunar landings and the other for space shuttle landings.

Countdown: Wednesday–Sunday 10:00 a.m.–5:00 p.m.; closed Monday and Tuesday

Mission budget: $

Flight team: 419-738-8811 | https://armstrongmuseum.org

Coordinates: 40.564, -84.1724 | 500 Apollo Drive, Wapakoneta OH 45895

Flight plan: The museum is a mile southeast of Wapakoneta, just off I-75 at exit 111.

Rations: Restaurants can be found throughout the town.

Orbital neighbors: Three hours west, on the campus of **Purdue University**, there's a bronze statue of alumnus Neil Armstrong, depicted as a college student, and a trail of Moon boot prints in front of the **Neil Armstrong Hall**

of Engineering. Flight team: https://engineering.purdue.edu/Engr/AboutUs /Facilities/ArmstrongHall.

Grissom Memorial at Spring Mill State Park, Mitchell IN

Debriefing: This small museum and memorial is a tribute to Indiana native Gus Grissom, who tragically died in the *Apollo 1* fire in 1967. Part of NASA's first three human spaceflight programs, Grissom was the second American man in space, on a suborbital flight aboard *Liberty Bell 7* for Project Mercury. He flew again for three orbits on the *Gemini 3* mission, and he was the commander of the ill-fated *Apollo 1* mission.

The STARS: The highlight here is the space-flown *Gemini 3* capsule that Grissom named Molly Brown. The choice of nickname came from the Broadway musical *The Unsinkable Molly Brown*, about a socialite who survived the sinking of the Titanic—a humorous reference to Grissom's previous Mercury mission in which the *Liberty Bell 7* capsule sank after landing.

Other artifacts include Grissom's *Gemini 3* space suit, plus personal and professional artifacts and clothing. Displays and scale models tell the story behind the three NASA programs Grissom participated in.

EVAs: The memorial is in Spring Mill State Park, which offers a variety of recreational activities, including camping, hiking, cave tours by boat, a nature center, and a restored pioneer village.

Countdown: Open daily 8:30 a.m.–4:00 p.m.

Mission budget: $

Flight team: 812-849-3534 | www.in.gov/dnr/state-parks/parks-lakes /spring-mill-state-park/

Coordinates: 38.728, -86.4137 | 3333 SR 60 East, Mitchell IN 47446

Flight plan: The park is about three miles east of Mitchell on IN-60

Rations: Restaurants can be found in nearby Mitchell.

Orbital neighbor: Located in the Golden Pond Visitor Center within Land Between the Lakes National Recreation Area in Kentucky, **Golden Pond Planetarium and Observatory** offers hosted star shows and nighttime telescope viewing. **Flight team:** https://landbetweenthelakes.us/seendo /attractions/planetarium.

Air Zoo Aerospace & Science Experience, Kalamazoo MI

Debriefing: The Air Zoo is an interactive museum that focuses mostly on aircraft, rides, and interactive activities plus some space offerings. Extensive aircraft galleries cover early flight to World War II, including a B-25 and an XP-55 prototype fighter from the 1940s. Jet age highlights include an SR-71 Blackbird, an F-18, and an F-117 Nighthawk stealth bomber.

The STARS: Highlights include exhibits spanning space history, including a Galileo display, simulated telescope, starfield, and presentation about science fiction classics. Artifacts include a primate capsule used by Ham the chimp in Project Mercury and a boilerplate Gemini capsule nicknamed El Kabong used for soft-landing drop tests for the canceled Para-Sail landing program. An *Apollo 11* exhibit recreates a suburban living room from 1969 where a family would have watched the historic landing. Nearby is a Moon rock from *Apollo 15*.

Other exhibits relate to the space shuttle program and International Space Station, alien worlds and androids, the contributions of women in aviation and space exploration, and the contributions of African American aviators and astronauts. In the future, the museum may add a Redstone rocket.

EVAs: Throughout the museum, there is a series of interactive stations, with some related to space. The museum has an extensive number of rides and flight simulators. The museum theater shows several films each day on various topics, including space.

Countdown: Monday–Saturday 9:00 a.m.–5:00 p.m.; Sunday 12:00–5:00 p.m.

Mission budget: $$

Flight team: 866-524-7966 | www.airzoo.org

Coordinates: 42.2275, -85.5582 | 6151 Portage Road, Portage, MI 49002

Flight plan: The Air Zoo is located just south of I-94, about midway between Chicago and Detroit.

Rations: The Kitty Hawk is the museum's on-site restaurant.

Orbital neighbors: Located in Bloomfield Hills MI, the **Cranbrook Institute** is a museum of natural history with a wide range of science exhibits, including some space-related offerings. There's also a planetarium and telescope observatory. **Flight team:** https://science.cranbrook.edu.

Museum of Science and Industry, Chicago IL

Debriefing: Located in waterfront Jackson Park, south of downtown Chicago, the Museum of Science and Industry is one of the largest science museums in the world. Over 400,000 square feet of exhibits range widely from the science of storms to melting Antarctic glaciers to human and animal genetics. There are recreations of a coal mine and an early twentieth-century Chicago neighborhood. A transportation zone includes a steam locomotive, the diesel train Pioneer Zephyr, a Boeing 727, a pair of World War II warplanes, and a replica *Wright Flyer*. One of the museum's top attractions is the U-505 submarine, the only German U-boat from World War II located in the United States. The sub was restored and moved inside the museum in 2004, where visitors can enjoy interactive exhibits and onboard tours, with advance tickets strongly advised.

The STARS: The Henry Crown Space Center at the Museum of Science and Industry offers a wide variety of highlights. Artifacts include the space-flown command module from *Apollo 8*, the first manned spacecraft to orbit the Moon. Nearby is Frank Borman's spacesuit from *Apollo 8* and a copy of the famous Earthrise photo signed by all three astronauts.

Placed within a simulated Moon landscape with mock-up astronauts on EVA, there's an authentic lunar module trainer used by all twelve astronauts who landed on the Moon. Other items include a collected Moon rock, space-flown lunar collection tools, samples of Apollo space food, and an authentic launch control panel from Kennedy Space Center.

Elsewhere, there's the space-flown Aurora 7 capsule from the second orbital mission in Project Mercury. Related to the space shuttle program, there's a full-size partial mock-up of the front third of an orbiter, a smaller scale model of the full stack, and an authentic unflown thermal tile. A mock-up International Space Station module includes interactive displays and a live feed. There is a simulated Mars surface where, at one time, visitors could remotely control a miniaturized rover that, during 2022, was being refurbished and may return to display in the future.

Two suspended satellites include a full-scale replica of Sputnik and the space-flown BioSat-2, which carried live organisms into space in 1967. Scale models reflect various space vehicles, including recent ones like the Ares I and Ares V from the canceled Constellation program, and a complete Falcon 9 with a Dragon spacecraft.

An exhibit called Spaceport presents the history and culture of space travel and rocketry. There's a full-scale diorama depicting Robert Goddard and his 1926 liquid-fueled rocket, the world's first, and a diorama about a V-2 rocket engine and White Sands. Other displays cover the Hubble and James Webb space telescopes. Science fiction items include uniforms from *Star Trek* and the "phone home" prop from *ET: The Extraterrestrial*.

Earth Revealed is a digital spherical project about Planet Earth that uses satellite and other data from NASA and NOAA to recreate weather formations, ocean currents, and seasonal variations.

Fast Forward is an exhibit focused on the innovative work of inventors and futurists. The featured innovators are periodically updated, with some recent projects related to space exploration, including the development of automated Mars rovers, food replicators, and more.

The gallery also includes a full-scale mock-up of the Mars *Curiosity* rover.

46. The *Apollo 11* lunar module floats in space before making its descent to the lunar surface. NASA.

The Giant Dome Theater has a five-story screen with rotating shows ranging widely in topics, from ecology to weather phenomena to space exploration. See the museum website for current offerings.

EVAs: The VR Transporter is an immersive virtual reality experience with recent first-person adventures in space travel, including the *Apollo* mission and a space shuttle space walk. Visit the museum website for current offerings. Interactive flight simulators allow participants to play the role of pilot or gunner in historic aircraft like a P-51, A-10, or F-15. Motion simulators are six-person motion theater pods allowing riders to experience a Moon base meteor storm, a high-speed outer space pursuit, stunt flying, or aerial combat. Tickets cost extra and can be purchased at the museum.

Countdown: Open daily 9:30 a.m.–4:00 p.m.

Mission budget: $$

Flight team: 773-684-1414 | www.msichicago.org

Coordinates: 41.7906, -87.5829 | 5700 S. DuSable, Lake Shore Drive, Chicago IL 60637

Flight plan: Paid parking is available in the museum's underground lot.

Mission parameters: The museum requires timed entry tickets and strongly suggests purchasing them in advance online.

Rations: The museum has several on-site eateries, including a restaurant, café, and snack bar.

Orbital neighbors: Located nearby on the campus of Triton College, the **Cernan Earth and Space Center** is a small museum and planetarium named for local NASA astronaut Gene Cernan. Cernan flew on *Gemini 9A*, *Apollo 10*, and *Apollo 17*, walking on the Moon during the final lunar mission. The center includes artifacts and exhibits like Cernan's *Apollo 10* spacesuit and undergarment, plus other Apollo-related items like navigation equipment and scale models. There's also a telescope exhibit and a reproduction of Copernicus's revolutionary 1543 manuscript arguing for heliocentrism. Outside the center, there's a Nike Tomahawk sounding rocket and an Apollo boilerplate capsule used by the U.S. Navy to practice ocean recoveries. The ninety-three-seat planetarium dome offers a variety of astronomy shows and space-related feature films for all ages. Visit the center website for ticket prices and showtimes. **Flight team:** www.triton.edu/cernan.

Adler Planetarium, Chicago IL

Debriefing: On the shore of Lake Michigan, within sight of the Museum of Science and Industry, the Adler Planetarium offers star shows, a telescope observatory, and exhibits related to astronomy and space exploration.

The STARS: The Astronomy in Culture exhibit explores stargazers and their tools throughout history, including the armillary sphere, astrolabe, sundial, and more. Telescopes: Through the Looking Glass includes a number of historic telescopes plus exhibits on the science behind the instrument.

Mission Moon presents the story of the Apollo program through the eyes of Captain Jim Lovell, NASA astronaut on *Gemini 7*, *Gemini 12*, *Apollo 8*, and *Apollo 13*. Artifacts include Lovell's lunar surface EVA helmet, a backup *Apollo 8* spacesuit, a portable life support system, and more.

Our Solar System includes artifacts and exhibits about the planets and Sun. There's a Moon rock, a meteorite that struck Arizona, a piece of Mars, and more. The Atwood Sphere shows the night sky as it appeared over Chicago in 1913, and Chicago's Night Sky focuses on constellations, celestial globes, star maps, and light pollution.

The Space Visualization Lab offers presentations by experts and interactive visualizations. The Community Design Labs and Play Labs are interactive exhibit spaces organized around different topics, including art, history, social science, and space travel. Planet X is a space-themed playground for children ages 3–8.

Planetarium sky shows and films in the dome theater are a major draw. They are shown throughout the day with most running about thirty minutes. In addition to astronomy shows, recent film topics have included the Moon, Pluto, and a simulated space tourism jaunt around the solar system. Tickets are available in advance online.

EVAs: Renovated in 2021, the Doane Observatory has a 24-inch reflector telescope available for public viewing on Wednesday nights, weather permitting.

Countdown: Thursday–Tuesday 9:00 a.m.–4:00 p.m.; Wednesday 4:00–10:00 p.m.

Mission budget: $$

Flight ream: 312-922-7827 | www.adlerplanetarium.org

Coordinates: 41.8664, -87.6072 | 1300 South DuSable Lake Shore Drive, Chicago IL 60605

Flight plan: Parking is available at an adjacent lot for a separate fee.

Mission parameters: The Adler Planetarium and Museum of Science and Industry are part of the Chicago CityPASS, which includes entry to five sites.

Rations: The **Cosmic Cafe** is the Adler's on-site restaurant.

St. Louis Science Center, St. Louis MO

Debriefing: Located in Forest Park, the St. Louis Science Center is a free interactive museum with a wide range of exhibits and artifacts on display. There are ten galleries with topics spanning dinosaurs, ecology, energy, engineering, and more. Space-related highlights include artifacts, replicas, and exhibits about the planets. There's also a four-story OMNIMAX theater and a planetarium. With a futuristic hyperboloid shape, the distinctive north entrance and planetarium were designed by famed architect Gyo Obata, who later designed the Air and Space Museum in Washington DC. A unique enclosed skybridge crosses I-64 to link the northern and southern halves of the museum.

The STARS: In the planetarium lobby, visitors will find some excellent artifacts in the Liftoff exhibit. The Mercury capsule #19 was the flightworthy backup for the *Mercury-Atlas 8* mission, flown by Wally Schirra in 1962. Gemini capsule 3A was the thirteenth flightworthy spacecraft built for the program, but never flown. Originally designed as a boilerplate test article, this capsule was later upgraded to a flight-qualified vehicle for heat and pressure testing. Nearby, there's a Mercury training suit that was used by astronaut Gordon Cooper. Rounding out this exhibition, there are several heat shields, both flown and unflown, an Apollo lunar surface TV camera, and some models.

Other exhibits focus on the planets, including Mars. There are full-scale replicas of three Mars rovers, *Sojourner* from 1997 and *Spirit* and *Opportunity* from 2004. A meteorite exhibit displays thirty samples from around the world, ranging from the very small to a 300-pound specimen found in Campo del Cielo, Argentina. This exhibit also includes a Martian meteorite.

While general admission is free, the center also includes a rotating series of ticketed attractions. These include special exhibitions, flight simulators, and a virtual reality experience. The OMNIMAX theater shows various films related to museum topics, with more information available on the center website.

Named for the founder of St. Louis-based McDonnell Aircraft, which built the Mercury and Gemini capsules, the James S. McDonnell Planetarium is a popular attraction at the center. Star shows are interactive events hosted by an astronomy expert using a ZEISS Universarium Mark IX, one of only three such optical star projectors in the country. (The others are at Griffith Observatory in Los Angeles and the Hayden Planetarium in New York.) Visit the center website for tickets and listings of upcoming shows.

EVAs: When visiting the center, consider checking out the rest of Forest Park. There are paths for walking and cycling and several free museums, including the St. Louis Art Museum, the Missouri History Museum, and the St. Louis Zoo.

Countdown: Monday, Thursday–Saturday 9:30 a.m.–4:30 p.m.; Sunday 11:00 a.m.–4:30 p.m.; closed Tuesday and Wednesday

Mission budget: Free

Flight team: 314-289-4400 | https://www.slsc.org

Coordinates: 38.6289, -90.2706 | 5050 Oakland Avenue, St. Louis MO 63110

Flight plan: There are two entrances, with the main entrance on Oakland Avenue just outside the park, where paid parking is located. Inside Forest Park, the northern entrance to McDonnell Planetarium once had free parking—this side closed during the COVID-19 pandemic but may reopen in the future.

Rations: There are several shops and food options inside the center, including the **Science Cafe and BBQ Grill**.

Orbital neighbors: The **St. Louis Air and Space Museum** has mostly aviation-focused displays with the exception of a few items, including two flight suits used by astronauts Gus Grissom and Gordon Cooper. **Flight team:** https://airandspacemuseum.org.

16 1969

The Pace Quickens

Early 1969 saw a frenzied pace of developments in the Apollo program. For one thing, the United States had a new president in the White House: Richard Nixon, who hung a framed copy of *Earthrise* on the wall in the Oval Office. But otherwise, it was uncertain how Nixon felt about the continuing Apollo missions.

NASA also saw a change in administrator. Previously appointed by Kennedy, James Webb was a lifelong government official who had overseen NASA since 1961. But when Johnson declined to seek reelection, James Webb resigned to allow deputy administrator Thomas O. Paine to become acting administrator. Paine was a materials engineer and nonpolitical space advocate, and the agency's hope that he would be retained came true when Nixon nominated him to the position.

Meanwhile, the lunar module (LM) was finally complete and ready for crewed testing. In early March, *Apollo 9* launched—with three astronauts—atop a Saturn V rocket, carrying the entire lunar spacecraft into orbit for the first time. After detaching from the S-IVB, the command and service module (CSM) turned around and docked with the LM, reverse thrusting to extract it from the third stage's payload bay.

Over eleven days, the comprehensive testing went perfectly. The astronauts propelled the combined craft by firing the main SPS (service propulsion system) engine and later the LM descent engine. One astronaut spacewalked from the LM hatch to the CSM hatch, proving it could be done in an emergency. They separated the lunar module from the CSM and flew the LM into a higher orbit, jettisoned the descent stage, and reunited with the CSM. The result was clear—the LM was ready for the Moon.

Two months later, *Apollo 10* launched toward the Moon in what was called a dress rehearsal for the first landing. From a stable lunar orbit, Ed Stafford

47. *Apollo 11* soars above the U.S. flag. NASA.

and Gene Cernan undocked in the LM and fired the descent engine, bringing them to about 50,000 feet from the lunar surface, the altitude from which powered descent would begin.

Due to the mission profile, and perhaps to avoid tempting the former test pilots, the LM was loaded with only enough fuel to orbit above the Moon but not enough to land. Floating over the Sea of Tranquility, the men took photos of the planned landing site for the next mission. Then they jettisoned the descent stage and used the ascent engine to reunite with the CSM. As they rose, an accidental computer entry caused the spacecraft to spin, but the crew regained control. On the return to Earth, *Apollo 10* set a speed record that still stands, traveling 24,791 mph. With another successful mission complete, the landing would proceed that summer.

Luna 15 Races to the Moon

On July 13, 1969, three days before the launch of *Apollo 11*, a Soviet Proton-K rocket leaped into the sky over Kazakhstan. Once in orbit, the Blok D upper stage initiated a translunar injection burn, sending a mysterious 12,000-pound spacecraft en route to the Moon.

Officially, the Soviets claimed that *Luna 15* was an automatic probe designed to conduct scientific explorations of the Moon. But the timing was certainly suspicious, and observers wondered about rumors the Soviets were trying to land a probe, collect a lunar rock sample, and return it to Earth. Doing so ahead of *Apollo 11* would give them the perfect opportunity to demonstrate the abilities of robotic spacecraft—and point out, by comparison, the inherent dangers and huge expenditures of the Apollo program.

NASA was confident that *Luna 15* was not an attempt at a manned landing. But the fear remained that *Luna 15* might disrupt radio communications with *Apollo 11*. Astronaut Frank Borman had recently returned from a goodwill visit with the Soviet space program. Sent by Nixon, Borman's purpose was to invite the Soviets to conduct a future joint mission with NASA. At the request of flight director Chris Kraft, Borman called Soviet space leader Mstislav Keldysh. The next day, Keldysh sent a telegram with the precise orbital trajectory of *Luna 15*, confirming it would not interfere with the *Apollo 11* mission. Over the course of the following days, *Luna 15* would lower its orbit in anticipation of landing. Rougher than anticipated topography at its intended landing site caused concerns that delayed the mission. When ground controllers resumed the descent of this sample-return probe, communication went silent, suggesting a crash into the lunar surface.

Apollo 11 Arrives

Throughout the morning of July 16, 1969, about a million people gathered near the Cape Kennedy launch site, filling beaches and highways. Watching from the NASA bleachers was former president Johnson, along with hundreds of government officials and about 3,500 reporters. Just after 1:30 p.m., the Saturn V thunderously arced into the sky over the Atlantic. Step by step, the mission proceeded like the run-ups before. Within hours of reaching low Earth orbit, the spacecraft began its three-day journey to the Moon. After a day of preparations in lunar orbit, Neil Armstrong and Buzz Aldrin entered the lunar module they'd named Eagle and separated from the command module named Columbia.

"The Eagle has wings," radioed Armstrong.

To initiate descent, the LM flew perpendicular to the lunar surface, close to a speed of 3,600 mph. With the descent engine pointed in the direction of travel, the astronauts executed a retrograde burn for thirty seconds. This slowed the LM down, allowing it to be pulled downward, by the Moon's gravity, into a lower orbit.

Inside the cabin, Armstrong and Aldrin stood in harnesses at the flight controls. Given the orientation of the spacecraft, they were essentially lying flat relative to the lunar surface. For now, the triangular windows located on either side of the control panel were pointed upward, out into space.

After coasting for an hour, the spacecraft reached an altitude of around 50,000 feet, where the powered descent began. During a sustained braking engine-burn of varying thrust levels, the spacecraft would take about twelve minutes to reach the surface. To start this burn, the spacecraft pointed its main engine at a slight upward angle, or pitch, to generate additional downward momentum. For the remainder of the burn, the craft would slowly alter its pitch toward a vertical orientation for landing.

About five minutes into powered descent, a series of distracting warning alarms sounded. Back on Earth, a frenzied mission control team confirmed these specific alarms didn't require an abort. The guidance computer was overwhelmed with incoming data, and it was prioritizing the most important tasks. The warnings would continue until landing, and the astronauts would have to ignore them.

A few minutes later, they slowed to a descent rate of around 70 mph as they neared an elevation of 5,000 feet. As the spacecraft pitched past 45°, both the forward momentum and descent rate continued to slow. They passed an altitude of 700 feet, dropping at about 20 mph, with the LM pitched back at about 30°. By now, the windows offered a view of the landing zone targeted by the guidance computer.

"Pretty rocky area," noted Armstrong. "I'm going to . . ."

Through the window, Armstrong saw the landing zone was a field of large

48. Neil Armstrong after landing on the Moon. NASA.

boulders, about six to ten feet in diameter. It was ejecta, resting around the near edge of the large West Crater.

Within a few seconds, Armstrong disengaged the autopilot and assumed manual control. He quickly pitched the spacecraft toward zero, an almost vertical orientation, which pointed the still-burning engine toward the ground. This decreased the descent rate and allowed the forward momentum to carry the LM over the gaping 600-foot crater. Traveling forward at a speed of 32 mph, the LM cleared the crater at a height of about 300 feet.

"Slow it up," said Aldrin.

Armstrong pitched the LM back to slow the forward momentum, but all this extra flying introduced a new issue.

"Okay, how's the fuel?"

49. Buzz Aldrin climbs down the ladder to the lunar surface. NASA.

"Eight percent."

"Looks like a good area here."

As Armstrong rocked the LM slightly forward and backward, Aldrin called out the height and speed in feet per second, while noting the fuel remaining.

"One hundred feet, three and a half down, nine forward. Five percent."

Armstrong didn't respond, fully focused on making an immediate landing.

"Picking up some dust," said Aldrin. "Thirty feet, two and a half down. Faint shadow."

Aldrin reported the contact light, which indicated the hanging wires on the landing legs had impacted the surface. Armstrong shut down the engine with less than thirty seconds of fuel remaining. The vehicle gently settled onto the surface. With engine shut-off, the plumes of dust ejecting from beneath the spacecraft instantly dissipated. The particles flew outward in ballistic sheets, continuing beyond the horizon and out of sight. The first strange phenomenon signaling their arrival on another world.

"Houston," radioed Armstrong. "Tranquility Base here. The Eagle has landed."

"Roger, Tranquility," said astronaut Charlie Duke from mission control. "We copy you on the ground. You got a bunch of guys about to turn blue. We're breathing again. Thanks a lot."

The flight plan called for a rest period, but the excited astronauts doubted they could sleep. Instead, preparations were made for the EVA (extravehicular activity). But between post-landing procedures and maneuvering in the cramped cabin, filled with checklists, food packets, and equipment, it took the two men longer than expected. Six and a half hours after landing, they were finally in their spacesuits and ready to emerge from the spacecraft. Armstrong climbed down the ladder until he stood on the footpad. As he stepped onto the surface, his boot sank into a thin layer of regolith.

"That's one small step for man, one giant leap for mankind," said Armstrong, becoming the first human in history to walk on another world.

"The surface is fine and powdery," continued Armstrong as he moved around. "I can kick it up loosely with my toe. It does adhere in fine layers, like powdered charcoal, to the sole and sides of my boots."

"Magnificent desolation," observed copilot Buzz Aldrin after exiting and pulling the LM hatch partially closed. "Making sure not to lock it on my way out."

The two astronauts spent about two hours and twenty minutes on the lunar surface. They took photos, performed experiments, and collected 45 pounds of rock samples for return to Earth. They found that walking in the one-sixth gravity worked best by loping, essentially taking long, almost hopping, strides. President Nixon called from the Oval Office to congratulate them.

At one point, Armstrong walked about 200 feet away from the LM to snap photos of a crater, while Aldrin extracted a few shallow core samples. Before departing, they erected a U.S. flag, which Aldrin saluted, and deployed a seismometer to measure moonquakes. After blasting off the surface, the ascent module successfully rendezvoused with the orbiting Columbia command module. And thus began the greatest disinfection program in space history.

50. Armstrong snapped this photo of Aldrin during their two-hour EVA. NASA.

17 They Came from Outer Space

A Threat from Lunar Viruses?

Throughout the 1960s, long before Armstrong and Aldrin walked on the Moon, there was a small but serious concern among scientists that Apollo astronauts might return to Earth with dangerous pathogens. A little-known scientist named Carl Sagan warned of lunar viruses that might decimate Earth biology. Or perhaps toxic molecules were hidden deep within Moon rocks. Other researchers were more worried about the spacecraft bringing Earth-based bacteria to the Moon, which would contaminate lunar samples and jeopardize experiments.

When NASA inquired, the U.S. Public Health Service bluntly stated that they would deny entry to returning lunar astronauts—which planet they'd be deported to was never specified—if the United States was at risk upon landing. The general public was all too familiar with this concept of extraterrestrial invasions, given science fiction efforts like Orson Welles's radio broadcast of *War of the Worlds* or Michael Crichton's recent novel *Andromeda Strain* about—hot topic—extraterrestrial microbes escaping from a crashed satellite and killing a town's population. Mix scientific caution with political necessity, and NASA developed a complex procedure to prevent contamination in both directions.

The spacecraft and equipment were assembled in clean rooms on Earth and sterilized using various combinations of chemicals, heat, and UV light. When cabin air was recycled or vented from the lunar module (LM) at Tranquility Base, it passed through bacterial filters. The collected rock samples were sealed into airtight stainless-steel briefcases that looked straight out of a James Bond film.

After the two spacecraft reunited in lunar orbit, the oxygen flow and venting systems were adjusted so the cabin air flowed from Columbia, carrying lunar particles on a steady breeze, into the LM, which was then jettisoned.

During the return transit to Earth, the three astronauts inside were regularly shoving a vacuum brush into every crevice of the cabin, hoping to suck up any remaining dust lingering on surfaces.

Welcome Back, Wear This

Nine days after leaving Earth, Columbia plunged heatshield first, like a flaming meteor, into the atmosphere. The parachutes opened and the craft splashed into the Pacific Ocean, 13 miles from the USS *Hornet* aircraft carrier. Recovery helicopters approached the floating capsule, dropping a decontamination swimmer in a wet suit, who affixed a flotation ring and rubber raft.

Through the open hatch, the recovery swimmers tossed inside biological isolation garments. A few minutes later, the astronauts climbed into the raft, wearing heavy nylon suits that covered them from fingertips to toes, complete with hoods, goggles, and respirators. After closing the hatch, the swimmer sprayed the capsule and its exhaust vents with iodine solution. Then the astronauts and swimmer sprayed each other, like a celebratory water fight with serious disinfectants.

During their helicopter flight, Collins and Armstrong became irritable from overheating in their suits. Meanwhile, the saltwater sloshing around their feet suggested the garments offered less than stellar biological isolation. Five minutes after boarding the USS *Hornet*, the astronauts were sealed inside a custom-built Airstream trailer, called the mobile quarantine facility. Watching through pressurized safety glass, they received a hero's welcome from President Nixon.

"Neil, Buzz, and Mike," began the president. "I want you to know that I think I'm the luckiest man in the world. And I say this not only because I have the honor to be president of the United States but particularly because I have the privilege of speaking for so many in welcoming you back to Earth."

That welcome included five days in their airtight trailer, beginning with a sea voyage to Hawaii—minus the ocean views. This allowed plenty of time to fill out their mandatory customs form.

- *Flight No.* Apollo 11
- *Origin*: Cape Kennedy
- *En Route Stop*: Moon

51. Shortly after splashdown, the *Apollo 11* astronauts emerged from the capsule wearing biological isolation garments. NASA.

- *Destination*: Honolulu
- *Cargo*: Moon Rocks and Moon Dust Samples
- *Persons on board known to be suffering from illness*: None
- *Any other conditions on board which may lead to the spread of disease*: To Be Determined

At Pearl Harbor, the Airstream trailer was lifted onto a truck. During a reception, the Honolulu mayor and thousands of residents celebrated the astronauts, like they were action heroes behind plastic. Next, a flight on a C-141 to Houston, where they waved to their wives. Another welcoming crowd, another flatbed truck. This one soon backed up to the recently completed Lunar Receiving Laboratory, a pressurized building that the crew entered through a plastic tunnel.

52. The three astronauts, inside the mobile quarantine facility, are greeted by President Nixon aboard the USS *Hornet*. NASA.

Now Things Get Weird?

During the following sixteen days, the astronauts stayed in the crew reception area while being tested by doctors and debriefed by the mission control team through bio-barriers. Meanwhile, a staff of twenty technicians and scientists—who had agreed to join the quarantine—went to work.

The Moon rock cases were opened and a small portion removed for experiments. In the run-up to the landing, this fraction, diverted for biological study, had become quite controversial. Most geologists, physical scientists, and NASA engineers believed the possibility of lunar life was incredibly remote, given the harsh surface conditions and constant bombardment by cosmic rays.

Why waste any of the limited samples returned? But any risks to the planet's biosphere, no matter how minor, were deemed too important to ignore. Throughout the Apollo program, everyone made sacrifices to get the job done.

Technicians stuck their hands into heavy gloves mounted inside vacuum chambers. They partitioned flecks of rock into petri dishes and test tubes, searching for microbes and toxins. If something deadly was discovered, the entire facility might have to be buried under a sarcophagus of dirt and cement, sacrificing the twenty-three people inside. Meanwhile, the staff and astronauts teamed up with a quarantine party for Neil Armstrong's thirty-ninth birthday.

As if things weren't strange enough already, next Moon dust was dumped like food flakes into fishbowls. Rock powder was fed to cockroaches and quails. Mice and oysters received lunar injections. The leaves of Earth plants were rubbed with exfoliating regolith treatments. Seeds from critical crops—like cabbage, tomatoes, onions, sorghum, and ferns—were planted in lunar soil. The staff watched and waited.

After two weeks the oysters were all, oddly, dead but everything and everyone else was doing just fine. The Moon plants were growing better and greener than the controls planted in regular Earth sand. Not only was the Moon—at least the part *Apollo 11* had visited—confirmed to be sterile, it seemed to have great gardening potential. Once that whole no-atmosphere challenge was worked out. The entire team was released from quarantine one day early.

Procedures were relaxed but mostly left in place for *Apollo 12*, which had a brief scare, during the 1969 holiday season, at the Lunar Receiving Laboratory. Pete Conrad and Alan Bean landed near the site of *Surveyor 3*, an unmanned lander that had been launched in 1967, and retrieved the probe's camera. Back on Earth, the Lunar Receiving Laboratory technicians discovered a small colony of common bacteria inside the lens. Was it possible? Life surviving for thirty months in a harsh vacuum that reached negative 250° Celsius? As it turned out, probably not. The technicians were the likely source of the bacteria, opening the camera in short sleeves while wearing only latex gloves.

The Moon was a great place to visit, but otherwise a rock devoid of life. The quarantine procedures were lifted after *Apollo 14*. Plenty of lessons had been learned, which might be helpful someday when astronauts came home from Mars.

53. *Apollo 12* astronaut Alan Bean inspects *Surveyor 3*, an unmanned probe that had soft-landed several years earlier. NASA.

18 The Landings Continue

A Long Step That Few Watched

After the success of *Apollo 11*, the next mission received far less attention. *Apollo 12* launched in mid-November 1969, with President Nixon in attendance. As the Saturn V streaked through a rainy sky, two bolts of lightning struck the vehicle. Luckily, there was no damage, and the rocket flew onward. On the ground, there was fear that the lightning may have damaged the explosive bolts that released the command module parachutes during descent. But since there was nothing that could be done, they didn't tell the astronauts and the mission continued.

Five days later, Conrad and Bean descended toward the surface in the lunar module, with a goal of nailing a pinpoint landing. They did so in the Ocean of Storms, only 500 feet away from the *Surveyor 3* probe, which had landed in 1967. "Whoopee!" said Conrad, as he stepped onto the lunar surface. "Man, that may have been a small one for Neil, but that's a long one for me."

At five-foot-six, he was almost five inches shorter than Armstrong. Plus, Conrad had made a bet with a reporter he could say anything he wanted upon first step. Not that many people back home were paying attention, especially after the color television camera broke.

The pair of astronauts completed two EVAs (extravehicular activities), with the second walk taking them on a long circular path that led toward the *Surveyor*. Along the way, they inspected craters, collected rock samples, and radioed observations back to Earth. To reach *Surveyor 3*, they descended partway into a shallow crater. Finding the probe intact on solid ground, they retrieved its television camera, soil sampler, and associated rock samples for return to Earth. All in all, it was a textbook mission. And during the now-common atmospheric descent, the parachutes opened just fine.

Houston, We've Had a Problem

By the time *Apollo 13* came around, in April 1970, landing on the Moon had become seemingly routine. Both the successful launch and early TV broadcasts during translunar injection received little coverage by the media. Command module pilot Jack Swigert was a late replacement from the backup crew. After he admitted on live TV that he forgot to file his tax return, and requested an extension from outer space, only mission controllers laughed.

But everything changed fifty-six hours into the mission, about three-quarters of the way to the Moon. Swigert flipped a switch for a routine stirring of the fuel cell oxygen tanks. The response was a loud bang, followed by dropping electrical power. The external thrusters began firing automatically, trying but failing to stabilize the craft.

"Houston, we've had a problem," said commander Jim Lovell.

Quickly, the ground team and astronauts went to work deciphering the situation. Seconds later, Lovell noticed, through the hatch window, they were venting a gas into space. In addition, small pieces of debris were floating around the spacecraft, suggesting an explosion. With power plummeting, the decision was made to shut down the damaged fuel cells and power down the command module. Simultaneously, lunar module pilot Fred Haise powered up the lunar module (LM), which the crew would use as a lifeboat. Not only was a lunar landing off the table, the goal now was to simply bring the astronauts home alive.

On Earth, the world was riveted by news broadcasts detailing the perilous situation. In space, after regaining attitude control, the question was how to proceed. Because the service module's main engine may have been damaged by the explosion, the astronauts used the LM ascent engine as a backup. A risky manually controlled burn was executed to put the spacecraft on a free-return trajectory where it would slingshot around the Moon and return to Earth. As the spacecraft swung around the far side, the crew set a record for furthest distance humans have ever traveled from Earth: 248,655 miles.

Through small windows, they watched the lunar surface pass beneath and recede behind them. Soon they initiated a second burn to speed up the return and land in the Pacific Ocean. But one complication was the debris still hovering outside their windows, obscuring star sightings used for navigational alignment. Using the Sun and Moon as fixed points, this manual burn was almost perfect. They shut down most of the LM systems to conserve supplies and waited out the voyage in a cramped LM cabin, which became increasingly cold and damp.

A new problem developed—rising carbon dioxide levels. The command module had plenty of scrubber canisters, but those were square while the LM system used round ones. During a hectic improvisation session at mission control, NASA engineers devised an adapter using materials aboard the spacecraft, including flight plan covers, plastic bags, and duct tape.

Leaping from challenge to challenge, the astronauts flirted with hypothermia in the unheated 38° cabin. Haise developed a urinary tract infection from reduced water intake. Because waste dumps might disrupt the trajectory, bags of urine accumulated in the cabin. Still, the spacecraft kept drifting off course, so two further manual burns were executed.

Approaching Earth, the service module was jettisoned, and the astronauts photographed the damage—a side panel had fully blown off, exposing damage all the way to the engine bell. Next, the LM lifeboat that had saved their lives was jettisoned. As the command module Odyssey began its fiery descent through the atmosphere, observers around the world held their breaths. After so many errant events, it was unknown if the heat shield was intact or if the parachutes would open. After landing safely in the Pacific, Lovell called it a successful failure. They hadn't landed on the Moon, but the three astronauts returned home alive. Once again, Apollo missions were suspended, and a subsequent investigation identified faulty wiring had caused the explosion.

The Final Missions

As early as December 1969, shortly after *Apollo 12*, President Nixon began questioning the need for further Moon shots. The program had succeeded in its Cold War objective. Additional landings now seemed superfluous from a political standpoint. The first mission to be eliminated was *Apollo 20*, followed by *19* and *18*. Thus ended any chance at a mission to the far side of the Moon. Production of the Saturn V was suspended after fifteen rockets were built. After the near disaster of *Apollo 13*, Nixon was calling for the cancelation of *Apollos 17* and *16*, as well.

The new NASA administrator, Thomas O. Paine, argued that these two final missions were extremely important, particularly from a scientific perspective.

NASA scientists believed they were on the verge of unraveling the origins of the Moon. Other considerations included the sudden unemployment of an entire industry, involving hundreds of thousands of space workers. Ultimately, the president agreed. Four more landings would proceed, with an emphasis on sites deemed of particular geologic interest.

In January 1971 *Apollo 14* landed near Cone Crater in the Fra Mauro formation, the first visit to the lunar highlands. This mountainous region was believed to be ejecta from a massive impact that formed Mare Imbrium. During the first of two surface EVAs, the astronauts got close to Cone Crater but failed to reach it due to disorientation and fatigue. Still, they collected many rock samples, including a large 20-pound boulder that, decades later, would be theorized to include an embedded meteorite from Earth. Standing near the LM, Commander Alan Shepard —the first U.S. astronaut in orbit ten years before—used a makeshift club to hit two golf balls across the lunar surface.

Late that July, *Apollo 15* made a hard but successful touchdown on a lava plain at the edge of Mare Imbrium, not far from Hadley Rille, a mysterious channel resembling a dry river valley. Landing partially in a small crater, the LM came to rest at a slight angle, luckily not enough to warrant an immediate abort. With three days on the surface, this mission was all about science. Before the astronauts removed their spacesuits, Commander David Scott opened the LM's top hatch and surveyed their field site.

The next morning, during the first of three EVAs, Scott and LM pilot James Irwin unloaded a new toy. The lunar roving vehicle was a 460-pound electric Moon buggy with a top speed of 8 mph. Cruising away from the LM, the rover crested small hills, and Hadley Rille came into view. Driving along the rim of this small valley, the astronauts observed large boulders spread across the slopes and the rille floor 1,200 feet below.

Scientists had long debated the formation process for lunar rilles, with some speculating they were created by flowing water. But the astronauts' photos and samples pointed toward a different story. With most collected rocks turning out to be basalt, a type of hardened lava, the origin of sinuous rilles was likely volcanic. Perhaps lava channels or formerly enclosed lava tubes, with the scattered boulders being remnants of the collapsed roofs.

The astronauts spent a total of 18.5 hours on the lunar surface. Outside the LM, they collected a core sample of lunar soil, and Scott performed an

54. The mysterious Hadley Rille can be seen behind astronaut David Scott and the lunar rover on *Apollo 15*. NASA.

experiment for the TV audience. Dropping a hammer and feather simultaneously, he demonstrated Galileo's theory that all objects, regardless of mass, fall at the same rate in a vacuum free of air friction.

While exploring nearby mountains, they visited the slopes of Mons Hadley Delta, where they collected a sample later named the Genesis Rock. Back on Earth, studies would date it to over four billion years old, possibly part of the Moon's primordial crust. The mission, however, was not without controversy, when it was later discovered the three astronauts were secretly paid by a private collector to carry postal covers to the Moon for later sale. The investigation derailed the astronauts' careers, and they never flew in space again.

For *Apollo 16*, the geologic surveys continued in the Descartes Highlands, a mountainous area suspected to be of volcanic origin. After the LM touched down, astronauts John Young and Charlie Duke emerged on the surface. Over three days, they used the lunar rover to visit craters, take photos, and collect samples that included the largest rock returned from the Moon. Called Big Muley, this breccia—a consolidation of rock fragments—weighed 26 pounds.

55. *Apollo 17* astronaut and geologist Harrison Schmitt surveys a large boulder at Taurus-Littrow. NASA.

For Duke's video camera, Young steered a "Grand Prix" drive that demonstrated the maneuverability of the lunar rover.

During their second EVA, they ascended 500 feet up the 20-degree slope of Stone Mountain, traveling 2.4 miles from the LM. During the third, they visited a large boulder called House Rock, which was peppered with holes from eons of micrometeorite strikes. Along the way, they were searching for volcanic rocks but never found any before blasting off the surface to return home. This meant the highland mountains were not volcanic in nature but created by impacts.

A Colorful Moon

For the final mission, in December 1972, *Apollo 17* sought out Taurus-Littrow, a valley in the Taurus mountain range, at the edge of the Sea of Serenity. Once again, the objective was investigating a new area of the lunar highlands, which satellite imagery suggested might be volcanic in nature. This time NASA had sent its first scientist to the Moon, a New Mexico geologist named Harrison Schmitt, who had been involved in astronaut training before becoming one himself. When he and Commander Gene Cernan stepped onto the lunar surface, four hours after landing, the primary goal was to search for volcanic rocks.

When one of them accidentally broke the rover fender, they tried a makeshift fix with maps and duct tape. Covered in kicked-up dust, they sampled craters, hills, and massive boulders. Along the way, they collected a breccia named the Friendship Rock, which would later be broken into pieces and distributed to all the nations on Earth.

Midway through the second EVA, the astronauts were exploring the rim of Shorty Crater when Schmitt spotted something unusual in the lunar soil. One constant from Apollo explorations, thus far, was that the Moon was essentially colorless—a dull gray rock in space. But at Schmitt's feet he saw faint but distinctly orange soil. In his excitement, the geologist couldn't hold his camera steady, and the resulting photos were blurry. But Cernan captured a clear shot of the orange phenomenon with a color scale.

Though it wasn't quite what NASA hoped for, the astronauts had succeeded. Later studies of Apollo samples and orbital imagery would reveal that very few mountains on the Moon were volcanic—most were created by impacts that excavated craters and flung out debris, creating elevated surface features.

But a clearer picture of lunar volcanic processes was developing, beyond the basaltic plains that created the dark lunar maria. When later analyzed, the orange soil contained tiny particles of volcanic glass. The eventual theory was a fire fountain. Billions of years ago, the Moon was a volcanically turbulent and magmatic place. Lava explosively burst from the low-gravity surface and rose high into the sky before raining down as tiny grains of orange glass.

"I'm on the surface," said Cernan, before reboarding the LM for the journey home. "And, as we leave the Moon at Taurus-Littrow, we leave as we came and, God willing, as we shall return, with peace and hope for all mankind."

Origins of the Moon?

During the decade following the Apollo missions, scientists pored over the collected samples and data. Among many curious discoveries, one was that the Moon has an almost identical oxygen-isotope ratio to the earth, suggesting a common origin. At first glance, this seemingly could support the theory that the earth and Moon formed from the same protoplanetary disk.

But radiometric dating of lunar samples suggested the Moon formed after the earth, around thirty million years later. Other observations related to the Moon's lower density compared to the earth, its relatively small core, and an absence of volatile elements like hydrogen, nitrogen, and carbon dioxide, which seemingly vaporized at some point in the past.

At a 1984 conference on lunar origins, the various observations were woven together into a radical new theory. About 4.56 billion years ago, the earth formed from the accretion of dust and rocky debris within a protoplanetary disk spinning around the Sun. At first the young Earth was partially molten and subjected to frequent impacts by other bodies tumbling through a chaotic early solar system. During its first thirty million years, the earth differentiated into a dense metallic inner core surrounded by a silicate mantle and a partly solid crust. Then a cataclysmic event occurred.

Coming from somewhere else in the solar system, a planetary body about the size of Mars smacked into the young Earth. The high-energy impact caused part of Earth's crust and mantle to shear off into space, while much of the impactor, called Theia, melted into the earth's interior. Over time, this molten debris orbiting the earth coalesced into a spherical Moon.

The theory wasn't perfect, but it did account for some of the peculiarities of the Earth-Moon system, including the high angular momentum—which could have been supplied by the impactor. Also, if the Moon formed from the relatively light materials of Earth's mantle and crust, that could explain the low lunar density and small metallic core. While variations of the theory continue developing to this day, the giant impact hypothesis soon became the prevailing theory of the Moon's formation.

Sites in the Great Plains
Five Adventures and Three Orbital Neighbors

Celestial Overview: Seemingly one of the smaller areas covered in this book, the Great Plains covers a lot of territory from south to north, as it's organized here. Within that region there are two must-see sites for any space enthusiast. In Kansas, the Cosmosphere has the largest exhibition of Soviet space artifacts outside of Moscow, plus tons of great NASA exhibits as well. Located in a small town in Oklahoma, the Stafford Air & Space Museum is an impressive facility displaying many space-flown artifacts and replicas. Elsewhere in the state, there are two orbital neighbors in Oklahoma City with worthy space exhibits to visit. And far to the north, the Strategic Air Command & Aerospace Museum is an impressive aviation facility with some space highlights to consider. Given this region is wedged between several other regions from this book, some Great Plains sites might be combined with trips elsewhere. For example, the Oklahoma sites might combine well with sites from bordering states New Mexico and Texas.

Stafford Air and Space Museum, Weatherford OK

Debriefing: One of the more impressive space exhibitions in this book is found in the small town of Weatherford on Interstate 40 in Oklahoma. Dedicated to NASA astronaut Lt. Gen. Tom Stafford, the 63,000-square-foot museum includes dozens of authentic and replica artifacts, including spacecraft, spacesuits, and related objects. Some artifacts relate to Stafford's missions—*Gemini 6A, Gemini 9A, Apollo 10,* and *Apollo-Soyuz*—while other exhibits round out the comprehensive displays.

Meanwhile, the aviation exhibits are equally impressive, with full-scale replicas of the 1903 *Wright Flyer* and earlier glider, Lindbergh's *Spirit of St. Louis,* and Yeager's *Glamorous Glennis* Bell-X1. Authentic artifacts include a T-38 Talon trainer, an F-117 Nighthawk stealth fighter, and many more.

The STARS: Outside the museum, visitors are greeted by a boilerplate Apollo capsule and several aircraft, including an A-10 Warthog. Inside, exhibits span the space age, including a full-scale replica of Goddard's first liquid-fueled rocket, from 1926; an authentic captured V-2 rocket engine;

56. A full-scale replica of the Apollo lunar module at the Stafford Air and Space Museum. Mike Bezemek.

and a full-scale replica of the Little Boy nuclear bomb, which was dropped on Hiroshima in 1945.

One highlight is the *Gemini 6A* capsule, flown by Stafford and Schirra in 1965 during the first space rendezvous with *Gemini 7*. Nearby, there is a full-scale replica of the entire *Gemini* spacecraft, including the equipment and retro modules. There's also Stafford's Project Gemini training suit and a flight backup astronaut maneuvering unit that went unused during *Gemini 9*. Among the largest artifacts is one of the few remaining flightworthy Titan II rockets, similar to those that launched the Gemini spacecraft, displayed horizontally in segments

From the lunar program, there's Stafford's flown space suit from *Apollo 10*, the capsule hatch door, and other small artifacts from the dress rehearsal mission prior to landing. There's a Moon rock from *Apollo 17*, an authentic console from mission control, and a tread shoe from the motorized crawler that transported Saturn V and Saturn IB rockets to the launchpad. Full-scale replicas include a lofted Apollo command and service module and a surface display with the lunar module and two astronauts on EVA (extravehicular activity). Related to *Apollo-Soyuz*, the joint mission between the United States and U.S.S.R., there's an authentic flightworthy backup docking ring.

A number of authentic rocket engines includes an F-1 from the Saturn V first stage and a J-2 from Saturn upper stages. Another unique item is an authentic NK-33, the first-stage engine from the canceled Soviet N-1 Moon rocket, one of only three such engines on display around the world.

From the space shuttle program, there's an authentic flown segment of a solid rocket booster, an equipment mounting pallet from Spacelab, and a flight deck simulator from Johnson Space Center. Rounding out the gallery is a wide range of scale models, including rockets from around the world and the Hubble Space Telescope.

EVAs: A small theater shows *The Wrong Brothers*, a ten-minute humorous documentary about the early days of aviation. Group tours for ten or more are possible with minimum two-week advance notice.

Countdown: Monday–Saturday 10:00 a.m.–5:00 p.m.; Sunday 1:00–5:00 p.m.

Mission Budget: $

Flight Team: 580-772-5871 | www.staffordmuseum.org

185

Coordinates: 35.5442, -98.6703 | 3000 Logan Road, Weatherford OK 73096

Flight Plan: The museum is just north of I-40, about a mile northeast of downtown Weatherford. Located on the historic U.S. Route 66, it makes an excellent stop on a long-distance or cross-country road trip.

Rations: A limited number of restaurants and lodgings can be found in Weatherford.

Science Museum Oklahoma, Oklahoma City OK

Debriefing: This massive 390,000-square-foot facility combines interactive science and engineering exhibits with historic artifacts and art installations. Hands-on children's exhibits include the Science Floor with an inside arena for Segway riding and other educational activities. A series of outdoor gardens includes exhibits about plants and butterflies, a greenhouse, and lawn games. The Aviation Gallery presents the evolution of flight, from Leonardo DaVinci's drawings to a replica *Wright Flyer* to a number of vintage aircraft. A comprehensive display of space offerings includes space-flown artifacts, full-scale mock-ups, space-themed exhibits, and a large planetarium.

The STARS: During 2022 the museum was preparing to renovate its astronomy hall and planetarium, with final exhibits subject to change. Check the website or call the museum for more information.

Two of the biggest space artifacts at the museum are a pair of engines from the Saturn V rocket. From the SI-C first stage, there's an F-1 engine displayed horizontally, which allows a close look at the components. From the S-IVB third stage there's a J-1 engine, capable of being reignited in space. From the Apollo service module, there is a flightworthy fuel cell of the type that exploded during *Apollo 13*.

A variety of spacesuits come from various NASA programs, including Gordon Cooper's Mercury suit and Tom Stafford's Apollo suit. Others come from Project Gemini and the air force's canceled Manned Orbiting Laboratory, an early concept for a crewed reconnaissance station.

Elsewhere, there's a display of meteorites, some of which can be touched, plus identical replicas of Moon rock specimens collected by Apollo astronauts.

There's an Apollo training simulator, a one-quarter-scale model of the lunar module, a mock-up Gemini capsule, and two mock-ups from Project Mercury, one of which is a partial simulator that kids can board.

Other items include a mock-up of Robert Goddard's liquid-fueled rocket, the first in the world. A full-size mock-up of the Nazi V-2 missile. A full-size model of Sputnik. And an LR-87 dual-engine unit from the Titan II, which launched Project Gemini spacecraft. A tire from the space shuttle program, and a toilet from the International Space Station. The new astronomy hall will have exhibits about the planets, including the search for exoplanets outside the solar system.

EVAs: Included with general admission, the Kirkpatrick Planetarium offers a variety of interactive star shows, presented live by a host who allows audience curiosity and input to guide this journey through the cosmos.

Countdown: Monday–Friday 9:00 a.m.–5:00 p.m.; Saturday 9:00 a.m.–6:00 p.m.; Sunday 11:00 a.m.–6:00 p.m.

Mission budget: $$

Flight team: 405-602-6664 | www.sciencemuseumok.org

Coordinates: 35.5238, -97.4755 | 2020 Remington Place, Oklahoma City OK, 73111

Flight plan: The museum is located just south of I-44 in northern Oklahoma City.

Rations: **Pavlov's Cafe** is the museum's on-site restaurant and snack bar.

Orbital neighbors:

Located in Oklahoma City, the **Oklahoma History Center** is a museum mostly focused on state history. The Launch to Landing exhibit commemorates the many Oklahomans who contributed to the space program, including *Skylab 4* astronaut William Pogue. On display is the space-flown Apollo command module from the *Skylab 4* mission, along with two Moon rocks, a display of flight suits from various astronauts, and other artifacts. **Flight team:** www.okhistory.org/historycenter.

Located 1.5 hours away, the **Tulsa Air & Space Museum and Planetarium** is a museum focused on regional aviation history with authentic aircraft and some space-related exhibits. Planetarium shows and films are available throughout the day. **Flight team:** https://www.tulsamuseum.org.

Cosmosphere Space Museum, Hutchison KS

Debriefing: The Cosmosphere combines the largest displays of Soviet and Russian space artifacts outside of Moscow with an impressive American exhibition. The result is a comprehensive space museum that should be on any enthusiast's list of must-see sites. In total, the museum houses about 14,000 items, of which only a fraction can be displayed at any one time, along with objects on loan from NASA and the Smithsonian. Since displayed objects do rotate, check the museum website to confirm current offerings.

The Cosmosphere also manages the off-site SpaceWorks. This highly regarded space artifact restoration and fabrication service regularly restores authentic spacecraft and produces high-quality replicas for film and TV, including the recent show *For All Mankind* on Apple TV+.

The STARS: Outside the museum, there's an authentic Mercury-Redstone missile with a replica spacecraft. An authentic Titan I ICBM has been repainted and topped with a replica spacecraft to resemble a Titan II from Project Gemini. Using a stairwell, visitors can descend below the Titan and examine the lower stage and engines.

Inside, the museum is organized into roughly eight galleries. The German Gallery focuses on the origins of the space age, including an authentic and refurbished V-2, with a cutaway panel revealing engine components, and an authentic V-1 Flying Bomb, an early pulsejet predecessor of the cruise missile.

The X-Plane gallery includes a movie replica of the *Glamorous Glennis*, the Bell X-1 rocket plane used in *The Right Stuff*. There's also a rocket engine from the subsequent X-15 program and Kansas astronaut Joe Engle's X-15 flight suit.

The Cold War gallery presents a variety of historical objects, including an authentic flight-ready backup of Sputnik and an engineering model of Sputnik II. There's a full-size replica of the first U.S. satellite, Explorer 1, and a flight-ready backup from Project Vanguard, which placed the second U.S. satellite in orbit. There's also a space-flown Jupiter nose cone, an Aerobee sounding rocket, and a segmented Redstone ICBM allowing closer inspection of the disabled warhead, nose section, and fuselage interior.

The Mollett Early Spaceflight Gallery includes plenty of highlights. The wreckage of the Mercury-Atlas 1 capsule, an unmanned test flight that crashed shortly after launch. Wally Schirra's space suit and an enclosed primate couch from Project Mercury. The space-flown capsule from *Gemini 10*, plus a space-suit used during the mission by Michael Collins. Soviet displays include a space-flown Vostok used as an unmanned satellite in the 1980s. A replica Voskhod depicted with a flight-ready Volga airlock used during the first human spacewalk, by Alexei Leonov in 1965. There are also cosmonaut spacesuits, including a space-flown suit worn by Svetlana Savitskaya in the 1980s, and examples of the SK1, Berkut, and Sokol. Another unique item is an authentic luna sphere, identical to the one on *Luna 1*, which became the first spacecraft to impact the Moon in 1959. Resembling a soccer ball, the sphere is made of many titanium pennants, small Soviet victory flags, which supposedly scattered across the lunar surface.

The highlight of the Apollo gallery is the space-flown command module Odyssey from *Apollo 13*, along with Jim Lovell's spacesuit. There's a Moon rock from *Apollo 11*, several space-flown cameras, including the Hasselblad that may have taken the iconic *Earthrise* photo, and film magazines from each mission. A simulated Moon display includes full-size replicas of the lunar module, lunar rover, and ALSEP experiments package, plus some Apollo training suits nearby. Visitors can enter one of the three original white rooms where astronauts prepared atop the service tower before launch, plus an authentic flight surgeon mission control console from Johnson Space Center. And a full-scale replica of the Apollo-Soyuz Test Project depicts the U.S. and Soviet spacecrafts docked as they were, in space, in 1975.

The Astronaut Experience gallery presents about a hundred artifacts from later space efforts like the Russian *Mir* space station, the NASA space shuttle program, and the International Space Station (ISS). These include the camera the ISS used to take the well-known photo showing smoke rising from the World Trade Center site in New York City on September 11, 2001.

EVAs: The Justice Planetarium offers star shows, typically twice a day, plus the dome theater offers documentaries and films. The interactive show *Dr. Goddard's Lab* explores the history of rocketry. Tickets are required and can be purchased in advance online.

Countdown: Wednesday, Thursday, and Sunday 9:00 a.m.–5:00 p.m.; Friday and Saturday 9:00 a.m.–7:00 p.m.

Mission budget: $$

Flight team: 800-397-0330 | https://cosmo.org/

Coordinates: 38.0655, -97.9212 | 1100 North Plum Street, Hutchinson KS 67501

Flight plan: The museum is located about a mile west of KS-61.

Rations: The **Cosmo Cafe** is the museum's on-site restaurant.

Strategic Air Command and Aerospace Museum, Ashland NE

Debriefing: Impressive aviation galleries are the main draw at this museum, including an SR-71 Blackbird, a B-52, a B-29, and a U-2 spy plane. But several space highlights make this a worthy stop for enthusiasts.

The STARS: Outside the museum, a rocket and missile park includes, among other objects, an Atlas ICBM and an SLV-1 Blue Scout, the first solid-fuel U.S. launch vehicle capable of orbiting a satellite.

Inside the museum, there are several items of significance. The restored Apollo CSM-009 command module was launched on a test flight in 1966,

57. With sleek lines seemingly straight out of science fiction, the SR-71 Blackbird (and its predecessor, the A-12) is a popular attraction at many air and space museums, including this one at Strategic Air Command and Aerospace Museum. Mike Bezemek.

atop a Saturn 1B rocket, to an altitude of 310 miles over the Atlantic Ocean. Nearby there's an Apollo boilerplate used for testing with a Saturn I rocket.

Rounding out the exhibition is an Atlantis space shuttle trainer, a Project VELA satellite from the 1950s through 1980s, which monitored compliance with the Nuclear Test Ban Treaty, and a prototype for an X-38, an unbuilt crew return vehicle for the ISS.

Other exhibits include Women in Aerospace, a display about the Tuskegee Airmen, and The Heartland Astronaut, about the career of Ashland native Clayton Anderson, who flew on two space shuttle missions and visited the ISS. Other space artifacts include flight suits, a flown shuttle tire, and cameras used by Gemini astronaut Wally Schirra.

EVAs: Public tours are given daily, typically starting at 11:00 a.m. and lasting two hours. Contact the museum to confirm.

Countdown: Open daily 9:00 a.m.–5:00 p.m.; closed major holidays

Mission budget: $$

Flight team: 404-944-3100 | https://sacmuseum.org/

Coordinates: 41.0177, -96.3198 | 28210 West Park Highway, Ashland NE, 68003

Flight plan: The museum is just off I-80 midway between Lincoln and Omaha, with free parking on-site

Rations: The SAC **Lunch Cafe** is the museum's on-site restaurant.

South Dakota Air and Space Museum, Box Elder SD

Debriefing: In western South Dakota, close to I-90, this museum makes a fine stop if you're visiting the area—possibly on a vacation to the Black Hills and Badlands National Park. The emphasis is aviation with several missiles and over thirty aircraft on display in an outdoor park, including a B-1B bomber, B-29 Superfortress, and T-38 Talon trainer. Inside the museum there are several more aircraft and a small space gallery with a few exhibits. Nearby, there's a quirky orbital neighbor with a single space exhibit partly made from wax.

The STARS: Outside the museum, there's a Titan I missile, which was the precursor to the Titan II, which launched the Gemini capsule into space. Inside the museum there's a small gallery, which includes model launch vehicles and spacecraft. There's an impressive display of false-color satellite imagery from the USGS Earth Resource Observation and Science program, while exhibits describe the program in detail.

Countdown: Tuesday–Saturday 9:00 a.m.–4:00 p.m.

Mission budget: Free

Flight team: 605-385-5189 | www.sdairandspacemuseum.com

Coordinates: 44.133, -103.0728 | 2890 Davis Drive, Building #5208, Ellsworth Air Force Base, SD, 57706

Flight plan: The museum is located outside the front gate of Ellsworth Air Force Base.

Rations: Restaurants can be found throughout the area, including nearby Rapid City.

Orbital neighbors: Located in the Black Hills town of Keystone, the **National Presidential Wax Museum** is about thirty minutes away by car. The museum has a life-size diorama of the 1969 scene aboard the USS *Hornet* with a wax President Nixon congratulating the three wax Apollo astronauts—Armstrong, Aldrin, and Collins—who are sealed inside the mobile quarantine unit. **Flight team:** www.presidentialwaxmuseum.com.

19 The End of the Space Race

After Apollo: Big Plans, Small Steps

As the final Apollo missions unfolded, NASA shifted focus toward what came next. In early 1970, the creator of the vaunted Saturn V, Wernher von Braun, moved from Huntsville AL to Washington DC to become the deputy associate administrator for planning, a position he held for two years before retiring out of frustration.

For years, the unofficial plan at NASA had followed the framework von Braun had laid out in the *Collier's* articles of the mid-1950s. Now von Braun, Mueller, and other NASA officials lobbied the Nixon administration to pursue a grand vision for future space endeavors.

For the short term, a permanently occupied space station in Earth orbit. To reach it, a low-cost reusable vehicle called the space shuttle. In the midterm, sending unmanned probes throughout the solar system, landing rovers on the Moon and Mars, and inventing reusable nuclear-powered rockets. As longer-term goals, a Moon base and crewed missions to Mars by the end of the century, perhaps as soon as the 1980s or 1990s.

Nixon and his advisers looked over these recommendations with skepticism. The Moon race had been won. Post-Apollo space travel need no longer be a priority, and NASA could compete for funding just like any other program. In essence, Nixon said, you can pick one big thing to do next. Since NASA couldn't build a permanent space station without a reliable way to reach it, only one program really made sense.

To gain wide approval, NASA pitched the shuttle as basically a space truck, a cost-effective way to launch all U.S. payloads. That much of the shuttle development would occur in California, a state Nixon hoped to carry in his 1972 reelection campaign, certainly didn't hurt.

Skylab Rises . . . and Falls

Once space shuttle development was underway, the question became what to do with the remaining Apollo hardware. Much would go to newly established space museums, especially engineering mock-ups, but other vehicles remained flight ready.

Ideas had long abounded, including the Saturn-Shuttle launch concept, with the shuttle being mounted to the upper stage of a Saturn V. Other concepts would use the Saturn V to launch a large lunar rover for a continued presence on the Moon. Further afield, a Saturn V might boost orbiters and landers to Mars, to prepare for an eventual crewed landing. But a Mars mission was off the table for now, as was the Moon base and permanent space station. Instead, the Apollo Applications program suggested converting an unused Saturn IV-B third stage into a smaller temporary orbital station. The result would be 82 feet long by 21 feet wide, weighing about 84 tons.

In May 1973 one final Saturn V roared off the launchpad at Kennedy Space Center. Once it was in the air, things went wrong almost immediately. A meteorite and sun shield were ripped off the station during ascent, along with one of two solar panels. Upon reaching orbit, *Skylab* was underpowered and overheating, with interior temperatures reaching 126° from direct sunlight.

This turned the first crewed visit into a repair mission. In May 1973 three astronauts launched atop a Saturn IB in an Apollo CSM. Once aboard *Skylab*, the astronauts deployed a reflective parasol to block the Sun. After temperatures dropped, the focus shifted to fixing the remaining solar panel. After twenty-eight days, they had stabilized the station and returned to Earth. Two more missions followed, the second staying for fifty-nine days and the third for eighty-four days.

During these visits, the astronauts gathered longer-duration experience with living and working in space. They slept in small upright quarters and used an innovative zero-gravity shower. They conducted experiments, most famously bringing spiders onboard, part of a student proposal, to test their web-spinning abilities in zero g. They used the Apollo Telescope Mount, a manned solar observatory, to take photos of the Sun at x-ray, ultraviolet, and visible wavelengths. That summer, they witnessed a solar prominence, an eruption from the solar surface of whiplike plasma filaments. And, in January, they watched Comet Kohoutek swing through the inner solar system.

58. *Skylab*, the first U.S. space station, orbits the Earth during 1973. NASA.

In February 1974 the final crew detached their command module for the return to Earth. *Skylab* was left in a stable and nearly circular orbit about 270 miles high. With consumables onboard for six more months of occupation, the plan was to return to America's first space station—maybe even expand it. NASA believed *Skylab* would stay up until the early 1980s, plenty of time for the new space shuttle to come online and boost the orbit. But in the following years, excessive solar activity heated the station and the uppermost layers of the atmosphere. The increased drag pulled the station into an increasingly low and unstable orbit.

In July 1979 *Skylab* reentered the atmosphere. During a fiery descent, the station broke apart into large pieces careening toward Earth. Around the world, the reentry had become a major media event, with the coverage seesawing between consternation and bemusement. Experts agreed, *Skylab* might hit anywhere. It was coming down ten years after *Apollo 11*'s great achievement. Clearly NASA could still build massive things, but could they control them?

Soothsayers made wild predictions about where the debris would strike. Gag products included *Skylab* repellent with a money-back guarantee. People held watch parties wearing helmets. The debris mostly fell in the Indian Ocean, but other intact pieces landed in sparsely populated areas of southwestern Australia. One item recovered was a 2,800-pound oxygen tank, and a local county fined NASA $400 for littering.

The final act of the Apollo program came in 1975. It began, curiously enough, with the launch of *Soyuz 19* from Kazakhstan in mid-July. A few hours later, a final Saturn IB and Apollo CSM launched from Florida. In low Earth orbit, the Apollo spacecraft extracted a small docking adapter and airlock. Over the course of a day, the Soviet and American spacecrafts maneuvered through orbit until rendezvous and docking.

Proceeding into the airlock, astronaut Ed Stafford and cosmonaut Alexei Leonov, the world's first spacewalker, united for a symbolic handshake in space. Broadcast live on television, the cooperative Apollo-Soyuz Test Project was more than just the first international space mission. Joined nose-to-nose, the silvery can-shaped Apollo and the green and white pepper-shaker-like Soyuz signaled a major political shift. After eighteen years, the Space Race was over.

The Cursed Planet

With a hiatus in crewed flights, and smaller budgets to work with, NASA spent much of the 1970s focusing on unmanned probes. Of particular interest was Earth's planetary neighbor, Mars. After several failures to reach the Red Planet in the 1960s, the Soviets had successfully soft-landed the first probe, *Mars 3*, in 1971, but it stopped transmitting after only fifteen seconds on the surface. More orbiters and landers followed, with some failing while others returned orbital photos. One Soviet probe parachuted toward the ground, sending several minutes of atmospheric data, but contact was lost shortly before landing.

Early American efforts during the 1960s brought a similar mix of failures

59. Utopia Planitia as seen from the *Viking 2* lander in 1976. NASA.

and successes. In 1965 *Mariner 4* observed a thin atmosphere and freezing temperatures of -150°F. Capturing the first close-up photos of another planet, the visit disproved a century-long theory that dark surface streaks were canals built by a Martian civilization. *Mariner 6* and *Mariner 7* executed equatorial and south polar flybys in 1969, returning a few hundred images depicting ice-filled craters and a polar ice cap.

Two years later, *Mariner 9* became the first spacecraft to orbit Mars. As a planet-wide dust storm subsided, photos revealed the massive Olympus Mons. Rising to an elevation of 72,000 feet, it is almost three times taller than Mount Everest on Earth. Covering 120,000 square miles, it is roughly the size of the entire state of New Mexico. Ultimately, this shield volcano proved to be the biggest and tallest mountain on any planet in the solar system.

In 1976 a pair of NASA Viking spacecraft entered orbit around the Red Planet. After over a month of surveying the surface for suitable sites, each probe dispatched a lander. A combination of heat shields, descent parachutes, and retro rockets brought these 1,300-pound crafts to successful touchdowns.

Viking 1 landed in rocky sand dunes of Chryse Planitia, east of Olympus Mons and the Tharsis volcanic plateau. *Viking 2* landed in Utopia Planitia, a vast northern impact basin with a rocky surface occasionally coated by a frost of water ice.

The squat landers carried about 200 pounds of scientific instruments, including a seismometer for detecting marsquakes and a surface sampler for analyzing the chemical properties of nearby soil. The most ambitious devices were for chemical and biological experiments, which included several different tests. Using the lander's robotic arm, soil samples were obtained and sealed inside test containers.

One test checked the sample for organic molecules, carbon-based compounds thought essential to life—none were found. Another test added nutrients and water to the soil and waited for signs of metabolism—once again, nothing. The next test exposed the sample to light, water, and an atmosphere of carbon dioxide and carbon monoxide, with hopes of stimulating photosynthesis. While a small amount of organic material seemed to be created, upon analyzing results, scientists concluded it was a nonbiological reaction.

Finally, during the test called Labeled Release, instruments applied a dilute nutrient solution, including radioactive carbon-14, and monitored the atmosphere above the sample for the production of radioactive carbon dioxide. The result was positive, with a steady emission of radioactive gases. The lead investigator insisted this was a sign of Martian organisms. But most other scientists disagreed, saying the overall results yet again suggested a chemical reaction, not life. The conclusions remained controversial, with decades of debate to come. Only one thing was certain: more landing missions were needed. But years would pass before the next probes arrived.

After the Viking Program, the rising costs of space shuttle development forced NASA to curtail Mars exploration. In the late 1980s, when the Soviet Union sent two probes to explore the Martian moons Phobos and Deimos, contact was lost—one while en route, the other in orbit. When NASA tried again in the early 1990s, communications with the *Mars Observer* went mysteriously silent three days before orbital insertion. A few years later, with the Soviet Union now dissolved, Russia launched *Mars 96*, but the Blok D upper stage burn failed, and the probe crashed back to Earth.

NASA finally scored two successful returns in the late 1990s. Landing with the sequential help of a parachute, retrorockets, and large airbags, Mars *Pathfinder* touched down in Ares Vallis. Satellite images of this long valley suggested it had been carved by flowing water. To investigate, the lander released a 23-pound wheeled rover named *Sojourner*, the first such vehicle on Mars. For 83 sols, a term for the 24-hour, 40-minute Martian days, the small rover made a slow 330-foot circuit partway around the rocky landing site. Along the way, it sampled a variety of volcanic and sedimentary rocks, which seemed to have been deposited by an ancient flood.

Meanwhile, above the planet, the Mars Global Surveyor slipped into a polar orbit. Using high-resolution cameras, the probe documented further signs that water had flowed across the surface, definitely in the past and possibly more recently. Other observations included the potential presence of recent lava flows, and a laser altimeter prepared a planet-wide topographic map.

The elation over these missions was short lived, however. As the new millennium approached, NASA saw more failures. The Mars Climate Orbiter lost communication during arrival due to an embarrassing design mix-up between metric and imperial units. And the Mars Polar Lander crashed near the planet's south pole. As the first forty years of Martian exploration came to a close, more than half of the missions had failed, earning Mars a new nickname: the cursed planet.

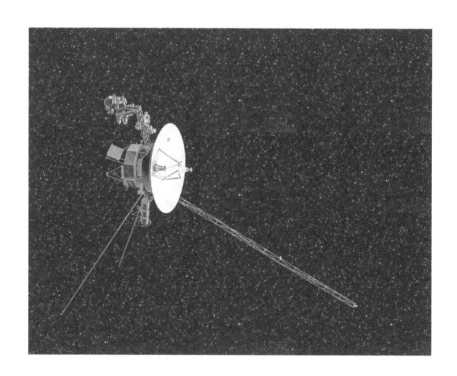

60. An artist's conception of the *Voyager* spacecraft. NASA.

20 Aiming for the Stars

The Grand Tour

During the 1960s, as post-Apollo space endeavors to the outer reaches of the solar system were contemplated, a NASA engineer named Gary Flandro made a compelling discovery. Every 176 years, each of the outer planets—gas giants Jupiter and Saturn, and ice giants Uranus and Neptune—lined up just right, on one side of the Sun, such that they could be visited by a single space probe. The next opportune alignment would serendipitously occur in the late 1970s, and a pair of complex spacecrafts named *Voyager 1* and *Voyager 2* were prepared for the occasion.

The key to what became known as the Grand Tour was to use a gravity assist at each planet to fling the probe onward to the next target. This type of gravity-assist maneuver would first be implemented on *Mariner 10*, which flew by Venus in February 1974, using the mysterious second planet's gravity to bend its trajectory toward innermost Mercury.

During the lead-up to the Grand Tour, two simpler probes, *Pioneer 10* and *Pioneer 11*, were launched toward Jupiter. *Pioneer 10* became the first spacecraft to visit the solar system's largest planet in December 1973. Coming within 80,000 miles, the probe returned helpful data about radiation levels and the magnetic field. Many low-resolution images were taken of the huge planet and its many moons, before the probe continued on a path toward the edge of the solar system.

Pioneer 11 reached Jupiter in December 1974, coming less than 27,000 miles from the swirling cloud tops and sending back the first images of the north polar region. Executing a slingshot maneuver toward Saturn, the spacecraft reached a top speed of over 107,000 mph before slowing due to gravity. It took five years to cross the 1.5-billion-mile distance.

Upon arrival, Saturn's first human-built visitor surveyed not only the planet but also its rings and moons. Observations of the largest natural satellite, Titan, confirmed it was the only moon in the solar system with a dense atmosphere, but at -330° F, it was likely too cold to support life.

With planetary explorations complete, the two Pioneer spacecraft continued their missions into deep space. Mounted onboard each probe was a curious plaque, promoted by astronomer Carl Sagan: the image of a nude human male and female, with pictorial representations of the probe's voyage through the solar system and the location of the Sun.

With lessons learned from the Pioneer probes, the two Voyagers flew away from Earth in the summer of 1977. At time of launch, space travel was a hot topic among the public. Audiences were swooning over the recently released *Star Wars*, an instant blockbuster that would fuel a rise of special effects–driven science fiction films. Compared to the two Pioneers, the superior Voyagers carried more powerful radioisotope thermal generators, remarkably sensitive cameras and instruments, and traveled faster than any human-made objects in history. But compared to the *Millennium Falcon* and *Death Star*, two small probes focused on factual science seemed downright academic.

Voyager 1 arrived at Jupiter in early 1979 and *Voyager 2* followed six months later. The images returned from throughout the Jovian system were astounding. The outermost Galilean moon, Callisto, was densely cratered. The next, Ganymede, was clearly the largest moon in the solar system, and its surface showed strange striations possibly tectonic in origin. Europa was crater-free and covered in ice. The innermost moon, Io, had active volcanoes, the first seen beyond Earth, seemingly caused by tidal stresses from the gravitational tug-of-war between Jupiter and the other moons. Around the gas giant itself, a faint ring system was detected for the first time, and a short movie made upon approach showed the planet's gaseous bands and storms rotating and twirling in various directions.

The next stop was Saturn, reached roughly twenty months later, where the two probes split up. *Voyager 1* made a hard turn toward Titan, coming within 4,000 miles. The planet's thick orange haze obscured the surface, but measurements of atmospheric composition, temperature, and pressure suggested the existence of hydrocarbon lakes. From there, *Voyager 1* passed under Saturn's south pole and departed the solar system's plane of the ecliptic, heading toward interstellar space.

Following nine months behind, *Voyager 2* sped over Saturn's north pole

61. *Voyager 1* captured this close-up image of Jupiter's Great Red Spot in 1979. NASA.

and swung onward toward Uranus. It became the first spacecraft to visit this icy blue giant in 1986. The discoveries were numerous: ten small new moons, two new rings, and intense radiation belts. The strange axial tilt of 97.8° meant the planet's axis of rotation is parallel to the ecliptic plane, unlike any other planet in the solar system. Possibly the result of an ancient collision. Most bizarre was its smallest and closest moon, Miranda. With the most extreme topography ever seen, the surface had racetrack-like grooves, rifts, and cliffs. One hypothesis was that Miranda had violently shattered in an impact before reforming into a jagged moon.

Voyager 2's final planetary destination was Neptune, reached in 1989. This ice giant was a darker hue of blue than Uranus, with Neptune's atmosphere having striations like Saturn. The probe passed only about 3,000 miles over the eighth planet's north pole, discovering wind speeds of over 1000 mph—the fastest in the solar system. Obtaining a more precise measurement of Neptune's mass, slightly less than thought, helped disprove theories that a hidden Planet X was responsible for minor peculiarities in the outer planet's orbits.

On its way out of the Neptunian system, the probe passed the largest moon, Triton, finding active geysers of nitrogen gas. The only large moon in the solar system with a retrograde orbit, Triton orbits opposite of the main planet's rotational direction. The theory became that the satellite was a captured object from the distant Kuiper Belt, a ring of comets and asteroids out near the orbit of Pluto. With *Voyager 2*'s planetary mission complete, it used Neptune's gravity to turn south out of the ecliptic plane and continue toward interstellar space and eventually out of the solar system.

Meanwhile, *Voyager 1* was now 3.8 billion miles from home, when a few scientists, including Carl Sagan, requested it look back for one final photo. Called *Pale Blue Dot*, a single pixel was caught in a beam of sunlight reflecting off the camera. Amid the vastness of space, the earth was just a pinprick of light.

From Concept to Commercial Shuttle

By the late 1970s, the space shuttle was taking form. The primary goal was to move away from single-launch rockets and disposable ballistic reentry capsules toward a reusable system that would drastically reduce the costs of accessing space. With the military on board as a partner, the shuttle would handle all U.S. space launches, perhaps as often as twice per month.

Early designs ranged widely. Some called for rocket-powered space planes, similar to the original Bell X-1 and North America's improved X-15, which could fly partway through the atmosphere toward space. Other early designs had the space plane launching on a piloted rocket booster that had its own wings, which then flew back down to land. Each concept had issues, mostly related to development costs, complexities, and payload limitations. Given that the U.S. Air Force was a stakeholder, they had mandated a 15-foot-by-60-foot cargo bay and lift capabilities of 65,000 pounds to low Earth orbit and 40,000 pounds to polar orbit. As costs rose, the Office of Management and Budget even stepped in, mandating the shuttle use cheaper solid rocket boosters instead of an entirely liquid fuel system.

The result was a partially reusable spacecraft. The delta-wing orbiter launched upright like a rocket, glided back down to a runway like a plane, and was readied to launch again like a commercial airliner. The main fuel

62. The space shuttle *Columbia* launches on its first test mission to low Earth orbit in 1981. NASA.

tank, which carried liquid oxygen and liquid hydrogen for the orbiter's three RS-25 engines, was disposable. Two side-mounted solid rocket boosters were jettisoned during launch and parachuted into the ocean, where they were retrieved and refurbished for future use. Altogether, the space shuttle was the most complex machine ever built.

Due to the integrated nature of the vehicle, the first launch was an all-up test flight flown by just two pilots. Commanded by *Apollo 16* astronaut John Young, with copilot Robert Crippen, the space shuttle *Columbia* launched on April 14, 1981. By coincidence, this was twenty years to the day after Yuri Gagarin's historic first spaceflight.

As *Columbia* arced through the sky, the astronauts sat strapped into ejection seats, modified from the high-speed SR-71 Blackbird. Though Crippen later questioned their effectiveness, given the ejecting astronauts would likely fall through the fire trail coming off the solid rocket boosters. Fortunately, they didn't have to find out. The mission was a perfect success. After a six-year hiatus since the Apollo-Soyuz Test Project, NASA was back in orbit with their fourth human spaceflight program.

During fifty-four hours in space, while orbiting the planet thirty-six times, the astronauts put the spacecraft through its paces. They fired the Orbital Maneuvering System, or OMS, a pair of hypergolic rockets mounted on the shuttle's tail. They measured radiation and tested the shuttle's star tracker. Then they opened the cargo bay doors to initiate the space radiators that cooled the shuttle's systems.

While doing so, they noticed damage to the thermal protection system tiles on the OMS pod. To protect the shuttle from the extreme heat of reentry, the orbiter was covered in several types of thermal tiles. Black tiles were used on the underside of the fuselage and wings, plus other areas exposed to extreme reentry heating, around 3000° F. White tiles were used where reentry temperatures were lower, typically around 1200° F. Placed across the upper portions of the fuselage and wing, these white tiles had the added benefit of reducing interior temperatures in direct sunlight.

Fortunately, the damaged tiles on the OMS pod were white, located in an area subjected to lower temperatures during reentry. Meanwhile, Young noticed two additional white tiles on the orbiter's nose with "big bites out of them." NASA had the astronauts maneuver the spacecraft in sight of a U.S. Air Force reconnaissance satellite. Ground controllers reviewed the photos, which showed the critical black tiles on the orbiter's underside were all intact.

Ultimately, reentry went perfectly, but the tile mishap emphasized the delicate nature of the shuttle's thermal protection system. Past NASA spacecraft had used ablative heat shields, which partially burned away during reentry and could not be reused. But the shuttle's large size and reusability demanded a lightweight and long-term solution. The over 20,000 ceramic tiles could repeatedly handle the heat, and be replaced as needed, but they were remarkably fragile. A tile might rip away due to aerodynamic forces during launch. Or the slightest impact, even from ice coming off the external fuel tank, could cause damage. Yes, the shuttle was the most advanced machine ever built, but its demanding design required that it operate at the very limits of current technology.

At the end of its test flight, *Columbia* glided down to a perfect landing at Edwards Air Force Base in California. A massive crowd awaited throughout the surrounding desert, cheering America's return to space. "This is the world's greatest all-electric flying machine," radioed Young. "I'll tell you that. That was super!"

Following three more successful test flights during the next fourteen months, President Reagan declared the vehicle operational. Now the space shuttle could fly with a full complement of seven astronauts, and NASA was free to start soliciting customers for its commercial launch program. The main competition was the Ariane Rocket, operated by the European Space Agency (ESA). After forming in the mid-1970s, the ten-member ESA focused on the growing commercial launch market, which Ariane began successfully serving in 1983.

That same year, the second space shuttle, *Challenger*, began flying. NASA released a glossy sales brochure titled "We Deliver." The pitch mentioned that NASA's launch record and experience meant the shuttle had "the lowest insurance rates in the free world." When launching a satellite, they claimed, you couldn't get a better bang for your buck.

New Astronauts and Opportunities

Another new development, since 1978, was an expanded astronaut corps open to female and minority applicants. When Eisenhower had limited astronaut selection to test pilots in the late 1950s, only white males had qualified. But civil rights legislation in the 1960s paved the way to new attitudes toward diversity in the armed forces and at NASA. Also, for the first time, shuttle missions would include astronauts other than pilots. Called mission specialists, now scientists, engineers, and even civilian guests would get an opportunity to go into space. With seven seats per mission, foreign astronauts from other space agencies would be eligible. Step by step, America's space program was becoming more representative of the diverse population across the nation and world.

On the seventh shuttle mission, Sally Ride became America's first woman in space, and third in the world after two female cosmonauts, Valentina

63. Sally Ride, aboard STS-7 in 1983, was the first American woman in space. NASA.

Tereshkova on *Vostok 6* in 1963 and Svetlana Savitskaya on *Salyut 7* in the early 1980s. The next flight included Guion Bluford, the first African American in space, the second person of African ancestry after Cuban cosmonaut Arnaldo Mendez. Soon came the first American woman to spacewalk. The first Canadian in space. The first Saudi. First Dutchman. First Mexican. First Japanese American. First sitting U.S. congressman.

A third space shuttle, named *Atlantis*, joined the fleet in 1985. Now operating a trio of orbiters, NASA was approaching a launch per month. Short clips on nightly TV news broadcasts had become routine, showing fiery but tidy launches, satellite deployments above a cloudy blue globe, and smiling astronauts floating in weightlessness. In fact, shuttle missions had become

64. Guion Bluford Jr., aboard STS-8 in 1983, was the first African American in space. NASA.

so routine that, like the latter Apollo missions, the American public began taking them for granted. To renew public interest, NASA began looking for new initiatives.

Ahead of the shuttle program's twenty-fifth flight, they unveiled the Teacher in Space project. From 11,500 applicants, a social studies teacher from New Hampshire named Christa McAuliffe was selected, and she began months of training with six other astronauts. She would fly as a payload specialist and offer televised lessons from space.

In late January of 1986, *Challenger* was ready on the pad, but poor weather and high winds delayed the launch for several days. Despite particularly cold temperatures on the morning of the twenty-eighth, the approval was given for launch. The bleachers at Kennedy Space Center were filled to capacity with spectators and family members. Classrooms around the country tuned in to watch.

As smoke billowed, the shuttle hurtled into the sky, trailing two plumes of fire from the solid rocket boosters. But seventy-three seconds after launch, there was a flash of light and *Challenger* was enveloped in a massive explosion. The side boosters broke free and rocketed away in different directions, as

65. An explosion during launch causes the space shuttle *Challenger* to break apart, killing all seven astronauts aboard. NASA.

trails of debris streamed down from the sky. The questionable ejection seats from the early test flights were long gone, and there was no way to escape. The orbiter was torn apart, and the crew cabin impacted the sea at over 200 mph. Observers were shocked, as the magnitude of this tragedy sunk in.

President Reagan canceled that night's State of the Union speech and delivered an address about the loss to a grieving nation. "We mourn seven heroes: Michael Smith, Dick Scobee, Judith Resnick, Ronald McNair, Ellison Onizuka, Gregory Jarvis, and Christa McAuliffe. We will never forget them, nor the last time we saw them this morning, as they prepared for their journey and waved goodbye and 'slipped the surly bonds of Earth' to 'touch the face of God.'" Five months later, the independent Rogers Commission submitted a detailed report on the accident. The fourteen-person investigation board included test pilot Chuck Yeager, astronauts Neil Armstrong and Sally Ride, and physicist Richard Feynman, among others.

They determined the accident was caused by the failure of a rubber O-ring, which had become brittle from the cold weather, located in a joint on the right-side solid rocket booster. Reviewing launch footage, fifty-nine seconds into the flight, a plume of flames was spotted shooting out of the joint. Like a blowtorch, the white-hot flame ruptured the external tank, igniting the propellants. Perhaps even more shocking was the revelation that some individuals within NASA and SRB contractor Morton Thiokol knew the O-rings were flawed and might fail in cold temperatures. In the run-up to launch, officials had debated the safety of proceeding but started the countdown anyway in hopes of keeping on schedule.

The commission made nine recommendations, and the shuttle was grounded for thirty months while these safety improvements were implemented. The SRBs were entirely redesigned. A limited crew escape system was implemented. The shuttle program was reorganized to allow any contractor or staff member to anonymously report safety concerns. Instead of increasing in frequency as originally planned, launches would be reduced. NASA would lose its monopoly on defense department payloads and exit the commercial launch market, ceding those services to unmanned rockets and private sector providers. When it returned in 1988, the shuttle would focus on scientific missions and building a permanent U.S. space station, which Reagan had approved several years before, named *Freedom*.

The Soviets Regroup

Lunar Rover Records

In the aftermath of losing the Moon race, the Soviets, in the early 1970s, needed to regroup. While much attention was paid to the series of mostly successful Apollo missions, the Soviets quietly continued their own unmanned lunar program. Possibly due to the death of Korolev, his massive N1 rocket had been a disappointment. All four test launches, spread from 1969 to 1972, resulted in failures. Three crashed into the ground after lift-off, and one exploded on the launchpad. A fifth was built but never launched, after the program was canceled.

But in September 1970, *Luna 16* succeeded where other efforts had failed. After landing in the Sea of Fertility, the probe deployed a drill, which extracted a 3.5-ounce rock sample from the lunar surface. With the sample secured inside, the ascent stage lifted off, and three days later it safely landed in Soviet territory.

Two months later, *Luna 17* landed in the Sea of Rains and deployed a remote-controlled rover onto the Moon. The first of its kind, the *Lunokhod* was the size of a small car and resembled a bathtub on wheels. The vehicle was powered by batteries and a round solar panel, which doubled as a protective lid that retained the warmth from a radioisotope heater during lunar nights. The rover traveled over six miles during 321 Earth days.

Though it received little attention, the Soviet Luna program continued until 1976, succeeding with two more automatic sample-return missions, two orbital surveyors, and one more rover. Landing inside Le Monnier crater in 1973, *Lunokhod 2* traveled 23 miles over four months, setting a record for off-Earth roving that eclipsed *Apollo 17* and lasted for almost four decades. The achievements of the Soviet lunar program were a welcome consolation, but for their next act in the dwindling Space Race, the focus shifted to a closer objective.

The Salyuts

Launched atop a Proton-K rocket in spring 1971, *Salyut 1* was the world's first space station. Orbiting about 130 miles high, the cylindrical station was 66 feet long and 13 feet wide, with room for three cosmonauts. A mission by *Soyuz 10* failed to properly dock, but a second visit by *Soyuz 11* successfully brought aboard the first crew. For twenty-three days, they performed experiments and conducted live television broadcasts.

At the end of June, the crew boarded their Soyuz for a return to Earth. But during separation of the reentry module, the explosive bolts misfired, which loosened a pressure valve, allowing cabin air to leak into space. When the ground team opened the capsule, the three cosmonauts were found dead from asphyxiation. These first human fatalities in space grounded the Soyuz fleet. Without a way to refuel the station, and given it was designed only as a temporary platform for short-term occupation, *Salyut 1* was purposefully deorbited several months later.

A year later, a second station was launched but it failed to reach orbit. Undeterred, the Soviets proceeded to a third iteration, naming it *Salyut 2*. But before its first occupants arrived, this one broke apart after only twenty-two days in orbit. A fourth attempt also met with failure, due to a launch malfunction that sent the station wildly spinning back into the atmosphere. To avoid embarrassment, the Soviets concealed the mission as a failed satellite.

After taking a year to regroup, the program finally rebounded in 1974 with a fifth station, officially called *Salyut 3*. Secretly, this station was actually a manned spy satellite called Almaz. During an initial sixteen-day mission, two cosmonauts operated a reconnaissance camera and conducted remote sensing experiments. Before being deorbited in early 1975, the station ejected a film capsule for recovery. Rumors even suggested the Almaz remotely fired an onboard gun into space.

Salyut 4 was a successful short-term civilian station that stayed up for just over a year. Two crews took advantage of many upgraded features, with one mission lasting for sixty-three days. *Salyut 5* was the final covert Almaz spy station, occupied by two crews who carried out reconnaissance for the Soviet military.

Two final Salyuts represented an updated second generation of Soviet space stations. Launched in 1977, *Salyut 6* had two docking ports, allowing dual missions to visit simultaneously. Now some cosmonauts would come for shorter visits, while others would stay for multi-month missions. A series of orbital marathons set impressive records for longest crewed missions in space: 96 days, 140 days, 175 days, and 185 days.

For their efforts, the Soviets were developing impressive credentials and making major strides in orbital science and logistics. To resupply the station, they developed the Soyuz-derived *Progress*, an uncrewed cargo ferry. After five years, the station was deorbited to make way for *Salyut 7*, an incremental improvement that would orbit from 1982 until 1991, pushing human duration records to 211 days and, later, 236 days.

Veneras to Venus

Elsewhere in the inner solar system, a Soviet fixation on Earth's nearest planetary neighbor was finally paying off. Since the early 1960s, the Venera program had sent a fleet of probes toward Venus. The planet is often described as Earth's twin, due to its comparable size and rocky mass, but that's where the similarities mostly end. With a dense atmosphere of primarily carbon dioxide creating a runaway greenhouse effect, the surface temperature is about 850° F, and the surface pressure is over ninety times that on Earth. Composed mostly of sulfuric acid droplets, the complex yellow cloud systems rise to heights of over 40 miles, completely obscuring the planet's surface.

After early flybys that lost communication, the Soviets' next success was with *Venera 3* in 1966—the first human-made object to impact another planet's surface. *Venera 4* repeated this success and returned atmospheric data, helpful for future missions. *Venera 5* and *Venera 6* were launched as similar atmospheric probes to pave the way for upgraded surface landers. In 1970 *Venera 7* became the first of these—heavily armored to survive the hellish conditions. Its parachute failed shortly before landing, and the probe toppled over but transmitted weak signals for about twenty-three minutes. Two years later, *Venera 8* made it to the surface and transmitted data for almost an hour.

The following Venera probes were much larger, each about five tons, launched from Earth using the Proton-K rocket. In 1975 *Venera 9* became the first spacecraft to orbit Venus. After separating from the orbiter, a 1,500-pound lander first used a parachute and then a heat shield to slow its descent through the dense fluidlike lower atmosphere. By the time it made a hard landing, it

66. A global mosaic of Venus created from imagery collected by the *Magellan* spacecraft in the 1990s. NASA/JPL.

was traveling only about 15 mph. After relaying the first photo taken on another planet, the spacecraft lasted only fifty-three minutes before succumbing.

The next three missions repeated this success, but issues with stuck lens caps meant only one more photo was obtained. Four additional Veneras rounded out the program. Two were landers, outfitted with microphones recording audio of a very slow and steady surface wind, basically a current of supercritical fluid. The other two were orbiters that used radar to image the surface.

By this time, the Soviets knew more about Venus than anyone. In 1986 they launched two more spacecraft, *Vega 1* and *Vega 2*. During flybys of Venus, each Vega spacecraft dispatched a lander, much like the previous *Venera* missions, but with a few new twists. After separation, the Vega motherships continued onward for a flyby of Halley's comet, becoming part of the Halley Armada, a half dozen probes—including Japan's *Suisei* and *Sakigake* and the European Space Agency's *Giotto*—sent to investigate the comet during its pass through the inner solar system.

And, for the Vega landers, at an altitude of 180,000 feet, each unleashed a small probe tethered to a 12-foot helium balloon. For at least forty-six hours, these small craft flew at 150 mph through the swirling clouds of Venus, tossed about by hurricane-force winds and traveling at least a quarter of the way around the mysterious second planet.

A *Mir* Revolution

Meanwhile, back in low Earth orbit, the final Salyut was not yet halfway into its life when the Soviet Union reinvented the nature of orbital space stations. In 1986 they launched the core module for *Mir* with an innovative plan to build the largest spacecraft ever. Amid the budget-conscious realities of post-Apollo space programs, von Braun's 1950s vision of a large wheel-shaped station revolving to create artificial gravity was simply unfeasible. The Salyuts and *Skylab* had shown that more realistic designs were based on relatively lightweight cylindrical modules, which could be most efficiently launched atop available rockets.

The *Mir* core module's revolutionary addition was a six-way node that would allow for modular expansion in six directions along all three axes. The result wasn't nearly as graceful as the pinwheel design. But a modular space station was much more practical, even if the awkward angles and jutting solar arrays created a chaotic appearance.

Over a fifteen-year life, the eventual seven-module *Mir* station would be continuously occupied for 3,644 days, including a still-standing record of 437 days by Valeri Polyakov. Mishaps would occur, including a small fire that filled the station with smoke and a collision with a *Progress* cargo ferry that damaged a module and sent the station tumbling. But each time, cosmonauts solved the problems. Meanwhile, microgravity experiments ranged across the disciplines, including astronomy, meteorology, physics, and biology. In addition to Soviet cosmonauts, visitors would include international and

67. The space shuttle *Endeavor* docks with the Russian *Mir* space station for the first time in 1995. NASA/JPL.

American astronauts. U.S. space shuttles would later visit, and Soviets even hoped to complete their own space plane to make the journey.

Buran: The Final Project

Called *Buran*, the Soviet space shuttle was authorized in the 1970s to match the capabilities of the U.S. program. Envisioned during the final decades of the Cold War, *Buran* was conceived as an answer to the American shuttle, which the Soviets feared would be used to militarize space, possibly by ferrying weapons, perhaps even lasers, to and from orbit.

Ready in the late 1980s, at first glance *Buran* appeared to be an exact copy of the American shuttle. Despite a black-and-white exterior with similar dimensions and characteristic delta wings, there were a few differences. The *Buran* orbiter carried no main engines for launch. They were located on the expendable super heavy-lift Energia rocket. The spacecraft was also configured for automatic uncrewed flight.

In 1988 the Energia rocket launched the first *Buran*. After it made two orbits, the guidance computer brought it down for a successful landing at Baikonur Cosmodrome. The biggest and most expensive Soviet space project had seemingly succeeded. But developments on the ground meant it would be the last space program of the U.S.S.R.

During recent decades, the Soviet economy had stagnated, mostly under Brezhnev and two brief leaders, Yuri Andropov and Konstantin Chernenko. In 1985 Mikhail Gorbachev became general secretary as a reformer, allowing increased freedom of speech and democratization and restructuring the economy to be more independent of the state. When discontented Soviet republics began seceding from the union, Gorbachev refused to intervene militarily. In 1991 the Soviet flag was lowered at the Kremlin and the Russian flag was raised. Boris Yeltsin became the first president of the Russian Republic. Rockets still flew from the Baikonur Cosmodrome to the *Mir* space station, but now they were Russian rockets flying from a leased spaceport in the independent nation of Kazakhstan.

Sites in the Southwest
Twenty-Six Adventures and Eight Orbital Neighbors

Celestial Overview: No need for a calculator; the numbers are correct. For being a somewhat smaller area—just four, albeit large, states—the Southwest accounts for more sites than any other region in this book. One reason is that the remote and unpopulated nature of the desert Southwest, particularly during the early days of the space age, was the perfect place for early rocket launches and atomic testing. Another reason is the desolate landscapes, with volcanic formations and craters, were excellent lunar analogs for astronaut training. Finally, high elevations, clear skies, and limited light pollution created perfect conditions for telescope observatories.

In fact, there are so many sites that the New Mexico Space Trail was established with over fifty space-themed stops, some more relevant than others. Eight of the best stops are featured in this book, with two must-see sites being the extensive Museum of Space History in Alamogordo and the nearby White Sands Missile Range Museum, where the U.S. space program got its start with captured V-2s and German rocket scientists.

Elsewhere in the region, look to Arizona for some unique highlights that offer a refreshing change from typical air and space museums. Meteor Crater is the best-preserved impact crater on Earth, also used for Apollo astronaut training. Plus, further astronaut training sites can be found around Flagstaff. You can also visit the famous Lowell Observatory, where Pluto was discovered in the 1920s.

Rounding out the region are about two dozen additional sites, with excellent museums, space artifacts, and observatories. There's a working spaceport and a famous telescope array. Some sites relate to atomic testing or weapons, while others are about UFOs. You won't get bored here, but you might run out of gas.

Mission parameters–NM Space Trail: The **New Mexico Museum of Space History** offers a detailed web page about each highlight on the New Mexico Space Trail. **Flight team:** www.nmspacemuseum.org/new-mexico-space-trail.

New Mexico Museum of Space History, Alamogordo NM

Debriefing: A comprehensive exhibition of historic space artifacts and replicas is spread across five floors in this quirky, cubelike building that feels straight out of the space age. There's also a planetarium dome theater and a rocket park outside with plenty to see, making this a top stop on the New Mexico Space Trail.

The STARS: In the John P. Stapp Air & Space Park, outside the museum, the biggest highlight is an 86-foot-tall Little Joe II. Five of these rockets were launched from White Sands Missile Range during the Apollo program to test capsule aborts using the launch-escape system. This is one of only two remaining examples of the Little Joe II, with the other on display in the rocket park at Space Center Houston.

Other space park items include an authentic F-1 engine from the Saturn V rocket, a rocket sled ridden by Stapp, a Thor rocket engine, and the grave of Ham, the first chimpanzee to fly in space. Nearby is the restored Daisy

68. A Little Joe II rocket from the Apollo program stands outside the New Mexico Museum of Space History. Mike Bezemek.

Decelerator, an air-powered sled-track used by NASA to study the effects of g-forces on the human body and equipment. Inside the Daisy Track exhibit building, there is a full-scale mock-up of the X-37, the U.S. Air Force/Space Force robotic orbital spaceplane. Rounding out the park is a boilerplate Apollo capsule used for sea recovery training, a series of sounding rockets and missiles, and other objects.

Inside the museum, five floors of exhibits present space history and artifacts. A rockets gallery combines authentic artifacts and components with replicas of engines, missiles, and more. There are full-scale replicas of the first Soviet satellite, Sputnik, the first U.S. satellite, Explorer, and an early U.S. guided missile, the Gargoyle. An authentic V-2 missile guidance system, and a space-flown guidance, navigation, and control system used on several Apollo missions. There's an Apollo lunar sample-return case and Moon rock.

A pair of artifacts relate to Ham, the chimp astronaut from Project Mercury who was trained at nearby Holloman Air Force Base, including a restraint suit and capsule couch. A number of spacesuits includes a replica from Project Mercury, a replica Soyuz Sokol suit used in the TV show *Big Bang Theory*, and others. Nearby, there are displays about space food, a space toilet, and a bio-instrumentation unit.

Other exhibits include a Friden Calculator, used during early NASA programs by human computers like Katherine Johnson, dramatized in the film *Hidden Figures*. There's an extensive gallery about the TV franchise *Star Trek*, including memorabilia, models, and props. The museum also includes the International Astronaut Hall of Fame, with displays on key figures from across the space age.

EVAs: The New Horizons Dome Theater & Planetarium offers live hosted star shows and films related to space and other topics. Visit the museum website for current offerings and showtimes, and tickets can be purchased at the museum. Also offered is a series of education programs for kids.

Countdown: Monday and Wednesday 10:00 a.m.–5:00 p.m.; Sunday 12:00–5:00 p.m.; Tuesday closed

Mission budget: $

Flight team: 575-437-2840 | www.nmspacemuseum.org.

Coordinates: 32.921, -105.9207 | 3198 State Route 2001, Alamogordo NM, 88310

Flight plan: The museum is on a hillside overlooking the northeastern corner of Alamogordo.

Rations: Restaurants can be found throughout Alamogordo.

Orbital Neighbor: About one hour from Alamogordo, in the remote Sacramento Mountains of Lincoln National Forest, the **Sunspot Astronomy & Visitors Center** is a solar telescope facility located at an elevation of 9,200 feet. The visitor center is open Thursday to Tuesday 9:00 a.m.–5:00 p.m. There is a museum with exhibits and, on select days and times, visitors can enter the observing room of the Dunn Solar Telescope. Contact the visitor center for more information. **Flight Team:** 575-434-7190 | https://sunspot.solar.

White Sands Missile Range Museum and Missile Park, White Sands NM

Debriefing: The White Sands Missile Range Museum and Missile Park presents the local origins of U.S. missile and space activity. Exhibits span the dawn of the atomic age, including the efforts of Wernher von Braun during the Space Race. An impressive outside missile park includes dozens of artifacts.

The STARS: The outdoor missile park displays about seventy missiles, rockets, and aircraft, many related to testing conducted at White Sands. Highlights include a Redstone ballistic missile, designed by Wernher von Braun for the U.S. Army in the 1950s, and the PEPP Aeroshell used by NASA to test the parachute for the 1976 Viking Mars lander. Other missiles include an Aerobee, a Pershing I, and a heat-seeking Sidewinder. There's a NASA control tower used during the 1982 test landing of space shuttle *Columbia* at White Sands, and a Fat Man bomb casing of the type used during the atomic bombing of Nagasaki in 1945.

Inside the museum, one highlight is an authentic V-2 captured in Germany after World War II, reassembled at White Sands, and restored in 2002 by the Cosmosphere. Other artifacts include a V-2 rocket motor and gyroscope.

During 2022 the museum was preparing to unveil updated exhibits about the history of the White Sands area, the role of the U.S. Army in early New Mexico, and the history of the White Sands Proving Ground and Missile Range. Check the museum website for updates on their offerings.

EVAs: In addition to walking through the rocket park, other outdoor activities can be found at the nearby White Sands National Park, where visitors can hike the stunning sand dunes or go dune sledding.

Countdown: Monday–Friday 8:00 a.m.–4:00 p.m.; Saturday 10:00 a.m.–2:00 p.m.; closed Sundays and holidays

Mission budget: Free

Flight team: 580-699-4798 | https://wsmrmuseum.com

Coordinates: 32.3852, -106.4785 | WSMR P Route 1, White Sands Missile Range NM, 88002

Flight plan: Visiting the museum requires obtaining a vehicle security pass. From Highway 70, take exit 172 for Owen Road. The Las Cruces/Alamogordo main post gate is approximately 3 miles after the exit. You can park in the visitor center parking lot and walk a short distance to the museum, in which case you need only a photo ID for admission. If you prefer to drive, park at the visitor center and take your driver's license, vehicle registration, and proof of insurance inside, where they will issue you a vehicle security pass.

Rations: The nearest city for restaurants and lodgings is Las Cruces NM.

Spaceport America, Truth or Consequences NM

Debriefing: Spaceport America is a commercial spaceport and active testing facility. Launch viewings are not allowed, and public access is restricted to private tours reserved in advance with Final Frontier Tours. These tours depart from the Spaceport America Visitor Center in nearby Truth or Consequences NM. Because of limited space, tours often sell out weeks to months in advance.

Spaceport America is the launch facility for Virgin Galactic's *SpaceShipTwo*, which began passenger flights in 2021, thus making New Mexico the third state in the United States to send people into space, after California and Florida.

The STARS: Tours typically last four hours and take visitors to the operations center, fire station, runway, and hangar where Virgin Galactic spaceplanes may be visible. Guests may have the opportunity to experience g-forces in a gyroscope trainer.

Countdown: Tours are available on Saturday and Sunday only

Mission budget: $$$$

Flight team: 575-267-8888 | www.spaceportamerica.com/

Coordinates: 33.1292, -107.2546 | 301 South Foch Street, Truth or Consequences NM, 87901

Flight plan: Spaceport America is closed to the general public except through guided tours from the visitor center in Truth or Consequences—do not try to drive directly to the spaceport.

Rations: Restaurants can be found around town before or after your tour.

Roswell Museum and International UFO Museum and Research Center, Roswell NM

Debriefing: In Roswell NM there are two small museums that may appeal to space enthusiasts. One is more traditional. The Roswell Museum has a gallery focused on rocket pioneer Robert Goddard, who lived and worked on rocketry in Roswell between 1930 and 1942.

Nearby, the International UFO Museum offers a fun change of pace from the mostly serious artifact-focused museums found throughout this book. The UFO museum combines informational displays with alien dioramas and a gift shop. The city is known for the infamous Roswell incident in 1947, when initial reports from the army indicated the recovery of a "flying disk." The claim was quickly retracted, and the recovered object was described as a crashed weather balloon. The incident led to decades of conspiracy theories about a crashed UFO and recovered extraterrestrials. The city has since become a hub for UFO-related attractions and events.

The STARS: The International UFO Museum presents facts and theories surrounding the 1947 incident. There are detailed dioramas of scenes related to conspiracy theories about the Roswell incident, including an alien landing and autopsy. Other displays present UFO incidents from around the world.

Countdown: Open daily 9:00 a.m.–5:00 p.m.

Mission Budget: $

Flight Team: www.roswellufomuseum.com

Coordinates: 33.3937, -104.523 | 114 North Main Street, Roswell, NM 88203

The STARS: The Roswell Museum is an art and history museum with several space-related offerings. One gallery is focused on Robert Goddard, displaying a recreation of his Roswell workshop, plus authentic rockets, components, tools, and other artifacts. The Goddard Planetarium is a digital dome theater offering shows that combine a film with a star presentation. Outside the museum, an outdoor exhibit combines the tower Goddard used to launch rockets with an example of one of his rockets and a statue depicting him observing the launch.

Countdown: Open daily 10:00 a.m.–6:00 p.m.

Mission Budget: $

Flight Team: 575-624-6744 | https://roswell-nm.gov/1259/Roswell-Museum

Coordinates: 1011 North Richardson Avenue, Roswell NM, 88201

Mission parameters: The two museums are less than a mile apart, either on or close to US-70.

Rations: Restaurants can be found throughout town.

Very Large Array, Socorro NM

Debriefing: On a high desert plain about 50 miles west of Socorro, the Very Large Array (VLA) includes twenty-eight sizable radio telescopes. When data is combined from each of the 82-foot dish antennae, the resulting resolution is equivalent to a telescope 22 miles in diameter. The VLA is one of the most recognizable astronomical observatories in the world, having appeared in many films and TV shows, including the 1997 movie *Contact* starring Jodie Foster.

The STARS: There are two main highlights here. The visitor center has exhibits about radio astronomy and a documentary narrated by Jodie Foster. The other highlight is a self-guided walking tour, which includes informational signs and leads to the base of one of the large dish antennae.

EVAS: Tours are offered on the first and third Saturdays, three times each day. Check the website for up-to-date information.

Countdown: Open daily 8:30 a.m.–sunset

Mission Budget: $

Flight Team: 575-835-7410 | www.vla.nrao.edu/

Coordinates: 34.0732, -107.6226

Flight Plan: From US-60, near mile marker 93, turn south onto NM-52 and proceed for 2.5 miles. Turn right onto NM-166 (Old Highway 60) and proceed 1.6 miles to the VLA visitor center.

Rations: Nothing. Maybe roadkill.

69. The Milky Way galaxy fills the sky above the Very Large Array in Socorro NM. NRAO/AUI/NSF/Jeff Hellerman.

New Mexico Museum of Natural History and Science, Albuquerque NM

Debriefing: An impressive natural history museum offering walk-through exhibits about the origins of life and the entirety of Earth history. There are several remarkably complete dinosaur skeletons plus several space-related galleries and a planetarium. Visit the museum website for an update on temporary exhibits.

The STARS: The museum displays a full-size replica of the 2003 Mars Exploration Rover (a.k.a. *Spirit* and *Opportunity*), which landed on the Red Planet in January of 2004. Over an increasingly extended mission of 5110 sols (Martian days), the latter rover traveled over 28 miles to the edge of Endeavor Crater.

227

Other museum highlights include a Moon rock collected in 1972 and a gallery focused on *Apollo 17* and the efforts of New Mexico astronaut and lunar geologist Harrison Schmitt. A small exhibit discusses the astronomical implications of Chaco Canyon.

EVAS: The museum planetarium offers a variety of space-related shows, including a history of human space flight. Check the museum website or call for current shows and times.

Countdown: Wednesday–Sunday 10:00 a.m.–4:00 p.m.

Mission budget: $

Flight team: 505-841-2800 | https://nmnaturalhistory.org

Coordinates: 35.0982, -106.6651 | 1801 Mountain Road NW, Albuquerque NM 87104

Flight plan: The museum is in downtown Albuquerque with free parking on-site.

Rations: The museum has a café on-site, and more restaurants can be found throughout the city.

National Museum of Nuclear Science & History, Albuquerque NM

Debriefing: Given the interwoven nature of the atomic age and the space age, this fascinating museum may be of interest to space enthusiasts exploring the New Mexico Space Trail. The extensive exhibits span atomic theory, nuclear weapons development, the bombings of Hiroshima and Nagasaki, nuclear politics, Cold War arms races, nuclear power, and more. Highlights include a B-52B Stratofortress and a B-29 Superfortress with the Fat Man bomb casing and transport container. There's a replica of the Trinity detonation tower, used in the first-ever explosion of a nuclear bomb in 1945. In addition, there are also several space-focused exhibits and artifacts.

The STARS: Heritage Park is a nine-acre outdoor area featuring the largest aircraft exhibition in New Mexico, along with rockets, missiles, and weapons. Large missiles include a Jupiter, Titan II, and Thor—three rockets that were also adapted for use as space launch vehicles. Other missiles include a Minuteman, A-3 Polaris, Matador, and more.

Inside the museum, exhibits include a chronological display presenting key discoveries and figures of the atomic age. The Critical Assembly installation is a recreation of the Manhattan Project's Los Alamos laboratories, as they appeared from 1943 to 1945, including authentic artifacts and replicas.

Decision to Drop presents the background of World War II, the secretive nature of the Trinity test site, and the reasoning behind dropping the atomic bombs on Japan. A Hiroshima and Nagasaki exhibit focuses on aftereffects on the people of the two bombed cities.

Cold War exhibits explore the nuclear arms race between the United States and U.S.S.R. from the 1950s to 1990s, with artifacts including bombs and missiles. Other exhibits focus on radiation, uranium and enrichment, energy, nuclear waste management, nuclear medicine, and changing attitudes toward the atomic age.

Countdown: Open daily 9:00 a.m.–5:00 p.m.; closed major holidays

Mission budget: $$

Flight team: 505-245-2137 | www.nuclearmuseum.org

Coordinates: 35.0664, -106.5335 | 601 Eubank Boulevard SE, Albuquerque, NM 87123

Flight plan: The museum is a few miles east of downtown Albuquerque, several blocks south of I-40.

Rations: Restaurants can be found throughout the area.

Space Foundation Discovery Center, Colorado Springs CO

Debriefing: This space, science, and technology museum combines interactive exhibits with a display of space artifacts. Colorado Springs is home to NORAD, the North American Aerospace Defense Command, a restricted military tracking center built deep inside Cheyenne Mountain.

The STARS: Found in the El Pomar Space Gallery, the Launch to the Moon exhibit focuses on early lunar exploration by both the U.S. and Soviet space programs. Items include a model Apollo lunar module and full-scale models of the Soviet *Luna* sample-return probe and the *Lunokhod* rovers. Other exhibits focus on uncrewed Moon exploration after Apollo, including the 2009 Lunar Reconnaissance Orbiter, the 2009 Lunar Crater Observation and Sensing Satellite, and the 2011 Gravity Recovery and Interior Laboratory.

There's a number of spacesuits including the 1980s NASA Extravehicular Mobility Unit and the Soviet Sokol K pressure suit. An exhibit on space food includes Russian and American food items from the space shuttle program and International Space Station. A spacelab exhibit allows visitors to explore the interior of the space shuttle orbital laboratory. There's a full-scale mock-up of a Viking lander, which touched down on Mars in 1976,

Several one-twentieth scale models represent the Ariane 44LP, Ariane 5, Atlas V, and Delta IV rockets. An interactive role-playing exhibit about a hypothetical mission to Europa, the icy moon of Jupiter. The Scott Carpenter Station, an underwater NASA research and demonstration vehicle from the late 1990s that simulated the isolated environment of space.

EVAs: The center offers a wide variety of programs and activities for school groups and youth, including a spherical projector used to depict the earth, Moon, Sun, and planets. Check the center website for current information.

Countdown: Tuesday–Saturday 10:00 a.m.–4:00 p.m.; closed Sunday and Monday

Mission Budget: $

Flight Team: 800-691-4000 | https://www.discoverspace.org/

Coordinates: 38.8957, -104.8633 | 4425 Arrowswest Drive, Colorado Springs CO, 80907

Flight plan: The center is found in northwest Colorado Springs, with free on-site parking, not far from Garden of the Gods.

Rations: The center doesn't have food, but many restaurants can be found throughout the surrounding city.

Orbital neighbor: Also in town, the **Peterson Air & Space Museum** is located on Peterson Space Force Base. Non-DOD members must request an appointment more than three days in advance of their visit. The free museum is mostly focused on aviation, but it has a few space-related items, including early satellites, plus some missiles and related artifacts. Appointments are possible Tuesday through Saturday 10:00 a.m.–3:00 p.m. For more information, visit https://petemuseum.org.

Wings Over the Rockies Air & Space Museum, Denver CO

Debriefing: Located in Denver, this impressive museum features over fifty aircraft and space vehicles, including a variety of unique space offerings. On the aircraft side, there's a B-52 Stratofortress, a Vietnam War–era Huey helicopter, and a rare B1-A Lancer, a test version of the supersonic bomber. The museum also offers several flight simulators, included with admission, while its off-site Blue Sky Aviation Gallery focuses on immersive experiences, discussed below.

The **STARS:** Perhaps the quirkiest item at the museum is a three-fourths-scale replica of the X-Wing Starfighter flown by Luke Skywalker in *Star Wars Episode IV: A New Hope*. Plus, there's a replica of Anakin Skywalker's pod racer from *Episode I: The Phantom Menace*.

Elsewhere, there's a mock-up of the HL-20 Dream Chaser, a prototype shuttle imagined by the Sierra Nevada Corporation to ferry astronauts to the International Space Station. Other space exhibits focus on the Titan missiles, which were produced in Colorado, displaying a first-stage engine, a second-stage engine, an interstage skirt, and informational displays. There's a Wings over the Rockies Air and Space Museum half-scale V-2 rocket replica and a variety of other artifact missiles.

There's a set of one-sixteenth models of the spacecraft from *Apollo 15*, plus a Moon rock from the actual mission. Rounding out the exhibition is an Apollo boilerplate capsule, an exhibit about Colorado astronauts, and a mock-up science module from the canceled space station *Freedom* project.

EVAs: Located at Centennial Airport, the Blue Sky Aviation Gallery is the museum's off-site immersive aviation facility with pilot-quality flight simulators and other interactive exhibits, including a gyro chair that simulates the effects of g-forces. Open Friday through Monday, the gallery requires the purchase of separate tickets, available on the main museum website.

Countdown: Monday–Saturday 10:00 a.m.–5:00 p.m.; Sunday 12:00–5:00 p.m.

Mission budget: $$

Flight team: 303-360-5360 | https://wingsmuseum.org

Coordinates: 39.7204, -104.8952 | 7711 East Academy Boulevard, Denver CO, 80230

Rations: Restaurants and stores can be found throughout the area.

Orbital neighbors:

Located at the Colorado School of Mines, about thirty minutes away in the foothills town of Golden, the free **Mines Museum** displays a Moon rock collected during the *Apollo 17* mission. Among the over two thousand items on display, there are minerals, fossils, rocks, and other objects related to the earth sciences.

On the campus of Denver University, the **Chamberlain Observatory** has public viewing nights hosted by the Denver Astronomical Society. Attendees must purchase tickets in advance. **Flight team:** https://science.du.edu/physics /chamberlin-observatory.

Space and Science Fiction Sites, Southern Utah

Debriefing: The otherworldly red rock country of Utah has often served as the surface of Mars and other planets in various science fiction films and TV shows. While most of the movie buff pilgrimages made to this region relate to Westerns, there are space-related sites to explore as well.

The STARS: Not only have parts of **Canyonlands National Park** been used in productions like HBO's *Westworld,* but one of the park's main attractions is potentially associated with meteoric activity. Located at the Island in the Sky district, Upheaval Dome is a mysterious crater that you can observe by hiking a short but steep one-mile trail. There are two competing theories about the creation of Upheaval Dome, an intrusive salt dome or a meteorite impact, with recent evidence pointing slightly toward the latter theory. **Flight team:** https://www.nps.gov/cany/index.htm.

With its fields of hoodoo rock towers, **Goblin Valley State Park** served as the surface of another planet in 1999's *Galaxy Quest* and has been featured in other films. Located near the town of Hanksville, the odd formations were named by nineteenth-century cowboys and can be explored up close by visitors today. **Flight team:** https://stateparks.utah.gov/parks/goblin-valley/.

The unpaved **San Rafael Swell Recreation Area**, administered by the Bureau of Land Management, was used as the planet Vulcan for the 2009 reboot of *Star Trek*. **Flight team:** www.blm.gov/visit/san-rafael-swell-recreation -area.

Monument Valley, located within the Navajo Nation and **Monument Valley Navajo Tribal Park**, was featured in several popular films, including *2001: A Space Odyssey* and *Back to the Future III*, and in many other productions. **Flight team**: https://navajonationparks.org/tribal-parks/monument-valley/.

Parts of Lake Powell in **Glen Canyon National Recreation Area** were used for the Martian river scenes in the 2012 big-budget flop *John Carter*—plus other locations were used around the state. Glen Canyon National Recreation Area was also used to depict the Forbidden Zone in *The Planet of the Apes* (1968).

EVAs: Though it is not open for public visitation or tours (despite third-party online info to the contrary), the Mars Desert Research Station does accept volunteers for their seasonal work parties and applications for crew members. The campus conducts field research and simulated Mars missions, typically in the range of two to three weeks. **Flight team**: http://mdrs.marssociety.org/.

Orbital Neighbors:

In downtown Salt Lake City, the **Clark Planetarium** offers star shows, IMAX films, and exhibits about Earth, the Moon, Jupiter's moon Io, the solar system, space weather, and more. **Flight team**: https://slco.org/clark-planetarium.

Near Provo UT, the **Christa McAuliffe Space Center** is a planetarium with star shows, films, and space mission group simulator programs. **Flight team**: https://spacecenter.alpineschools.org.

Meteor Crater, Winslow AZ

Debriefing: Fifty thousand years ago, an iron-nickel meteor about 150 feet across plummeted through the atmosphere at around 26,000 mph. When it impacted the ground in present-day Arizona, it excavated a crater three-quarters of a mile wide and 700 feet deep.

Today, Meteor Crater is the best-preserved and most-studied impact site on Earth, which makes it a must-see for many space enthusiasts. Visitors can observe the crater from several viewpoints, watch a film about its formation, and learn about meteorites in the Barringer Space Museum.

The STARS: By far the highlight of any visit is viewing the crater from the museum grounds on the north rim. There are several viewpoints reached by paved paths. A short walk along the rim is possible by joining a guided trip (included with the price of admission) during which you can also learn more about the history of the crater.

Inside the museum there is a fifteen-minute movie about the crater formation, while the Barringer Space Museum focuses on the crater's origin story;

70. This isn't the Moon from orbit but an aerial photo of Meteor Crater in Arizona, believed to be the best-preserved impact crater on Earth. USGS.

information about meteors, comets, and asteroids; and other impact features around the planet. Just outside the museum hall is the Holsinger Meteorite, the largest discovered fragment of the 150-foot meteor that created Meteor Crater. In the museum courtyard, an Apollo boilerplate capsule is on display.

EVAS: For more views of the crater, join a guided rim tour, included with general admission.

Countdown: Open daily 7:00 a.m.–7:00 p.m.

Mission Budget: $$ (purchase tickets online to save a few dollars)

Flight Team: 928-289-5898 | https://meteorcrater.com/

Coordinates: 35.0331, -111.0217 | Interstate 40, Exit 233, Winslow AZ, 86047

Flight Plan: Meteor Crater is a thirty-minute drive from Winslow AZ. From I-40 take exit 233 and drive about 6 miles south on Meteor Crater Road.

Rations: There is an on-site restaurant, the **Blasted Bistro**. More restaurants and stores can be found in Winslow AZ.

Lowell Observatory, Flagstaff AZ

Debriefing: Located in the hills near Flagstaff, the Lowell Observatory describes itself as the Home of Pluto, a reference to the facility's role in the discovery of what became the solar system's ninth planet. In the early twentieth century, Percival Lowell extensively searched—to no avail—for the mysterious and elusive Planet X. Unbeknownst to Lowell, he captured two photographs of the faint dwarf planet in 1915, but they went unrecognized during his lifetime. Fifteen years later, a young astronomer at the Lowell Observatory, named Clyde Tombaugh, captured proof of tiny Pluto orbiting in the far reaches of the solar system.

The STARS: Highlights at the observatory include exhibits and artifacts with nighttime opportunities for telescope observations, and tickets include all-day in-and-out privileges. The Rotunda Museum features exhibits about the discovery of Pluto, mapping the Moon during Apollo, expansion of the universe, Percival Lowell's early research about Mars, and more.

The famous Pluto Discovery Telescope is open briefly during the day and most evenings with general admission, as is the Clark Refractor, a 24-inch telescope that is open most evenings. The Giovale Open Deck Observatory is a public observing plaza with daytime exhibits and six advanced telescopes available during nighttime star parties for public use.

Scheduled to open in 2024, the Astronomy Discovery Center will include the unique Universe Theater, with wrap-around and overhead LED screens for films and live shows; a rooftop Dark Sky Planetarium with open-air nighttime presentations and heated seats; an interactive Curiosity Zone for children; and an exhibit hall organized around the story of atoms from the Big Bang to modern life.

EVAs: Included with general admission, several scheduled science talks and constellation tours are held each evening during telescope viewing hours. Every Saturday evening, the observatory has a scheduled Meet an Astronomer presentation with a Lowell researcher. For a premium fee, the observatory offers private seventy-five-minute group access to the 24-inch Dyer Telescope.

Countdown: Wednesday–Monday 10:00 a.m.–10:00 p.m.; Tuesday 10:00 a.m.–5:00 p.m.; Memorial Day to Labor Day 10:00 a.m.–11:00 p.m.

Mission budget: $$$

Flight team: 928-774-3358 | https://lowell.edu

Coordinates: 35.2028, -111.6647 | 1400 West Mars Hill Road, Flagstaff AZ, 86001

Flight plan: The observatory is located about a mile west of downtown Flagstaff.

Mission parameters: Located at an elevation of 7,000 feet, nighttime temperatures in Flagstaff can be quite cold, even in the summer. Check the weather forecast ahead of time and come prepared.

Rations: Flagstaff is filled with many great restaurants, microbreweries, lodgings, and stores.

Astronaut Training Sites, Northern Arizona

Debriefing: During the Apollo program of the 1960s, in anticipation of landing on the Moon, several places around Flagstaff and Northern Arizona were used as astronaut training sites. In addition to Meteor Crater and Lowell Observatory, discussed above, there were several smaller sites that can still be visited. These sites are atypical compared to the standard space museums covered in this book. Two of the sites are National Park Service units where astronaut training occurred, while one site is now an off-highway vehicle recreation area, and another is a USGS astrogeology center.

The STARS: At **Sunset Crater Volcano National Monument**, outside of Flagstaff, the Bonita lava flow was used for training several astronauts, including Gene Cernan and Jack Schmitt. They came in 1972 ahead of the *Apollo 17* mission, which hoped to find further evidence of volcanism on the Moon. Around the jagged lava flow and cinder fields, they tested equipment, practiced geological survey techniques, and used Grover, a lunar rover test vehicle. Today, this small monument is an excellent way to explore one of the main terrestrial landscapes chosen to simulate the lunar environment. Visitors can hike around the lava flow, cinder field, and volcanic cones on a series of short trails. **Flight team**: www.nps.gov/sucr

On the outskirts of Flagstaff, the area now known as **Cinder Hills OHV Area** has a fascinating history related to the space age. In the 1960s, NASA and the USGS used explosives in a volcanic cinder field to create two sets of craters for training Apollo astronauts.

The northern site, Crater Field 2, was about 1,200 square feet and, today, only faint outlines remain at what has become an off-roading area in Coconino National Forest. This site is accessible by unpaved Forest Service Road 76, off of US-89.

The southern site, Crater Field 1, was about 500 square feet and constructed to replicate the anticipated landing site for *Apollo 11*, including a mock-up lunar module. Today, this field is behind a chain-link fence designed to keep OHV riders from disturbing the craters. There are small openings in the fence so visitors can explore the site by foot. From US-89, follow Cinder Lake Landfill Road northeast for about two miles. Park outside the landfill and hike one mile east across the cinder field toward a large clearing, where you will find the craters. **Flight team**: https://www.fs.usda.gov/recarea/coconino/recarea/?recid=70996.

The **Astrogeology Science Center** in Flagstaff is a USGS facility founded in 1963 to map the Moon and help NASA conduct Apollo astronaut training. Today their efforts relate to exploration of the solar system, planetary mapping, and more. During COVID-19, the center suspended public tours, but these may resume in the future. **Flight team:** https://www.usgs.gov/centers /astrogeology-science-center.

About an hour and a half drive north of Flagstaff, the South Rim of **Grand Canyon National Park** is a must-see sight for many people. During astronaut training for *Apollo 11*, Neil Armstrong, Buzz Aldrin, and the backup crew hiked through the canyon with scientists to study geology. They hiked down on the South Kaibab Trail, spent a night at Phantom Ranch, and hiked out on the Bright Angel Trail. This classic trip is physically demanding and quite popular, requiring participants to be in excellent physical condition and acquire the necessary reservations ahead of time. Visit the park service website to learn more. **Flight team:** www.nps.gov/grca/index.ht.

Mission parameters: Flagstaff AZ makes for an excellent base of operations to explore these astronaut training sites. There are many hotels, motels, restaurants, and other sites, like Lowell Observatory and nearby Meteor Crater.

Pima Air and Space Museum, Tucson AZ

Debriefing: With around four hundred historic aircraft, ranging from a full-size *Wright Flyer* replica to a B-29 Super Fortress, an SR-71 Blackbird, and a prototype Boeing 787 Dreamliner, the Pima Air and Space Museum is one of the largest private aviation museums in the world. In addition to the impressive aviation artifacts, the small aerospace gallery is worth a stop, and a likely highlight for space enthusiasts is NASA's Super Guppy.

The STARS: Outside the museum, a Thiokol solid rocket booster from the space shuttle program is on display. In the space gallery, visitors will find several items of interest. A replica of the 2007–8 Phoenix Mars lander. A full-scale Mercury capsule trainer. An Apollo command module mock-up, originally built for the CBS *Evening News* and later used by Ron Howard and Tom Hanks in the miniseries *From the Earth to the Moon.*

Located in the outdoor aviation park, the Super Guppy is one of the space highlights of the museum. Five of these aircraft, considered some of the most unique ever built, were constructed by Aero Spacelines beginning in the 1960s. The enlarged cargo area, with a diameter of around 25 feet, was designed to transport oversized items like the Saturn IV-B stage.

EVAs: The museum offers two tours for an additional fee. The **Tram Tour** takes guests on a forty-five-minute tour of over 150 aircraft throughout the museum's eighty-acre site. The **"Boneyard" Tour** takes guests onto Davis-Monthan Air Force Base to tour the over four thousand aircraft that are part of the 309th Aerospace Maintenance and Regeneration Group.

Countdown: June–September: 7 days a week, 9:00 a.m.–3:00 p.m., with the last admission at 1:30 p.m.; October–May: 7 days a week, 9:00 a.m.–5:00 p.m., with last admission at 3:00 p.m.

Mission Budget: $$

Flight team: 520-574-0462; https://pimaair.org/

Coordinates: 32.1399, -110.8646; 6000 East Valencia Road, Tucson AZ, 85756

Flight plan: Located near the junction of I-10 and I-19 in Southern Arizona, the museum is 10 miles southeast of Tucson AZ.

Mission parameters: With much of the museum located outdoors in the hot Arizona Sun, make sure to bring sunscreen and drink plenty of water.

Rations: The museum's Flight Grill restaurant is located on-site.

Orbital neighbors: There are two small university museums nearby that may be of interest to space enthusiasts.

Located on the University of Arizona campus, in downtown Tucson, the **Flandrau Science Museum & Planetarium** has a solar system gallery with scale models of the planets, a gallery of high-resolution satellite images of Mars, and other science exhibits. A telescope observatory is open to the public a few nights each month. The planetarium offers star shows and films in a domed theater. **Flight Team:** https://flandrau.org.

Located southwest of Tucson, the **Kitt Peak National Observatory** offers several nighttime observing programs related to astronomy, deep-sky objects, and the Moon. Daytime tours are possible. **Flight Team:** https://visitkittpeak .org.

Located on the Arizona State University campus in Tempe, about two hours' drive from Tucson, the **Buseck Center for Meteorite Studies** has a small meteorite gallery that's free and open to the public. **Flight Team:** https://meteorites.asu.edu/visit.

Titan Missile Museum, Green Valley AZ

Debriefing: Located about thirty minutes south of Tucson, the Titan Missile Museum preserves the last remaining Titan II underground missile site in the United States, with an inert Titan II missile in its silo. Operational from 1963 to roughly 1987, fifty-four Titan II silos were located in four states—Arizona, Arkansas, California, and Kansas. The Titan II was an intercontinental ballistic missile (ICBM) developed for the U.S. Air Force to be used as a second-strike response in the event of nuclear war.

In addition to its use as an ICBM, the Titan II was adapted into a space launch vehicle for government payloads and all twelve Project Gemini missions. The focus here is on the Titan II's use as a Cold War ICBM. But when considering its relevance as a space vehicle and the museum's proximity to Pima Air and Space Museum, this stop may appeal to some space enthusiasts.

The STARS: Space enthusiasts may enjoy the opportunity to inspect the Titan II up close from underground level two inside the missile silo. The tour involves descending 35 feet into the access portal, inspecting the blast doors and walls that protected the launch crew from nuclear detonations, visiting the launch control center, and following the cableway tunnels into the silo.

EVAS: In addition to the guided tour, visitors can explore the topside of the missile site, with highlights that include a ground-level view down into the silo.

Countdown: October–May: daily 9:45 a.m.–5:00 p.m.; June–September: Thursday–Monday 9:45 a.m.–5:00 p.m.

Mission budget: $$

Flight team: 520-625-7736 | https://titanmissilemuseum.org

Coordinates: 31.9027, -110.999 | 1580 West Duval Mine Road, Green Valley AZ, 85614

Flight plan: The museum is about 20 miles south of Tucson, west of I-19, on Duval Mine Road (exit 69).

Mission parameters: Visiting the silo requires joining a forty-five-minute walking tour that involves descending and ascending fifty-five steel grate steps. Space is limited on each tour, so online reservations are strongly encouraged.

Rations: Plenty of restaurants and stores can be found nearby, with many more available in Tucson.

National Atomic Testing Museum and Nevada
National Security Site, Las Vegas NV

Debriefing: Las Vegas offers access to a pair of sites related to atomic weapons testing that may appeal to space enthusiasts. The National Atomic Testing (NAT) Museum offers exhibits and artifacts related to nuclear science and history. The museum also tells the story of the U.S. nuclear weapons testing program at the Nevada National Security Site (NNSS), a separate area involved in astronaut training that can be toured by reservation only.

Formerly known as the Nevada Proving Grounds or Nevada Test Site, the NNSS is a U.S. Department of Energy research complex covering almost 1,400 square miles, about 65 miles northwest of Las Vegas. The site offers day-long public tours every month by reservation only, for those at least fourteen years old. The tours typically depart from the NAT Museum by chartered bus, covering about 250 miles, and visitors must wear long pants and sturdy shoes.

The STARS: The highlights at the NAT Museum include history exhibits that span the atomic age, from World War II through the Cold War and into the present era. A particular focus is on the many nuclear tests conducted at the NNSS.

Artifacts include a nuclear reactor used to develop early nuclear rockets and air-to-air missiles, a series of decommissioned and replica bombs and missiles, a backpack nuke and the Davy Crockett recoilless nuclear gun, a piece of the destroyed Berlin Wall, two pieces of the wreckage from the World Trade Center, memorabilia reflecting nuclear age culture, and a replica of the Control Point that conducted countdowns for each test. The Ground Zero Theater simulates an above-ground nuclear test.

EVAS: Free public tours of the NNSS are offered by reservation only, on a monthly basis, for ages fourteen and up. These tours typically depart from the NAT Museum in Las Vegas around 7:30 a.m., by charter bus, and return around 4:00 p.m. The tour covers 250 driving miles, plus walking on rugged terrain, and long pants and sturdy shoes are required. Participants can bring their own food or purchase lunch at the site bistro.

Highlights of the tour include a visit to the mostly abandoned on-site town of Mercury, the Frenchman Flat detonation site, the hazardous materials testing complex, the radioactive waste management complex, the nuclear incident exercise site, the Icecap tower from a canceled 1993 underground test, and the Apple 2 test houses that were subjected to atomic blasts.

One of the most popular stops also carries significance for space enthusiasts. Sedan Crater was created in 1962 by a 104-kiloton thermonuclear explosion, with the resulting crater being just under 1,300 feet wide and 320 feet deep. During the Apollo program, astronaut training was conducted, at Sedan Crater, on the geology of impact craters that might be encountered on the Moon. **Flight team:** https://www.nnss.gov/pages/PublicAffairsOutreach /nnsstours.html

Countdown: Thursday–Tuesday 9:30 a.m.–3:00 p.m. (last entry)

Mission budget: $$

Flight team: 702-409-7366 | https://nationalatomictestingmuseum.org

Coordinates: 36.1143, -115.1484 | 755 East Flamingo Road, Las Vegas NV, 89119

Flight plan: The museum is located a mile east of the Las Vegas strip. The NNSS can be visited only through tours reserved in advance.

Rations: Restaurants can be found throughout Las Vegas.

Extraterrestrial Highway, Highway 375 NV

Debriefing: Different from most adventures included in this book, the Extraterrestrial Highway is an alien-themed road trip route in central Nevada running for about 100 miles on NV-375 from Tonopah in the northwest to Alamo in the southeast.

The existence of the ET Hwy is primarily due to its proximity to Area 51, the secretive U.S. Air Force facility that wasn't officially acknowledged until 2013. The facility's activities are classified as top secret, probably because the air force is developing experimental aircraft and weapons there. Over the years, Area 51 has been associated with many conspiracy theories and alleged UFO sightings. Note that Area 51 is heavily guarded and not open to the public.

There are no museums or artifacts to visit along this route related to actual space travel, but there are some campy tourist attractions that may be of interest to space enthusiasts. Because of the dark skies, the area is a great place for stargazing. A short detour leads to Lunar Crater, a volcanic crater used for astronaut training for *Apollo 17*. Put it all together, and it's a bit like a shorter UFO-themed Route 66.

The STARS: The town of Hiko includes several popular stops, often for photo ops, including the famous highway sign and the Alien Research Center, with its two-story alien sculpture. The inside houses a gift shop with a wide range of alien merchandise.

The town of Rachel has a few highlights. It's the closest town to Area 51, and therefore a popular spot for UFO enthusiasts. The Little A'Le'Inn is a small inn with a campground, bar, gift shop, and advice for UFO spotters. About 20 miles southeast of Rachel, a popular stop is the mysterious black mailbox.

In Alamo, E. T. Fresh Jerky offers exactly what it sounds like. The Nevada National Security Site is also near the highway, with monthly tours available and discussed in the other NV adventure chapter.

EVAs: For adventurous road trippers looking to leave the pavement behind, a detour north from the ET Hwy on US-6 and dirt roads leads to Lunar Crater. This free site, managed by the Bureau of Land Management, is an undeveloped national natural landmark, named for its resemblance to impact craters on the Moon. However, the half-mile-wide and 400-foot deep Lunar Crater was created by earthly volcanic eruptions. In the 1970s, the crater was used as a training site for astronauts on *Apollo 17*, in hopes they might encounter evidence of volcanism on the Moon.

Countdown: Most visitors will aim for spring through fall to drive the ET Hwy.

Mission budget: Driving the highway is free, other than fuel costs.

Flight team: https://travelnevada.com/road-trips/extraterrestrial-highway/

Flight plan: You can drive the highway in either direction and, given its proximity to Las Vegas, this can be combined with a space-themed road trip that includes California and Nevada sites.

Rations: Other than a small restaurant at the Little A'Le'Inn and snacks at ET Fresh Jerky, there is not much to find on the highway itself. There are restaurants and stores in Tonopah.

71. Astronauts from the space shuttle *Discovery* service the Hubble in orbit during STS-103 in 1999. NASA/JSC.

Adventures in Orbit

Hubble Discoveries

With upgrades, the U.S. space shuttle program resumed operations in 1988. Two years later *Discovery* launched the highly touted Hubble Space Telescope (HST). When its main mirror was found to be flawed, a new orbiter named *Endeavor* was sent up, in 1992, and the crew performed a daring repair job over 300 miles above the planet. Now that it was fixed, and operating well outside Earth's disruptive atmosphere, the HST images were stunning. Millions of spiral galaxies filled the cosmos, along with colorful nebula and celestial nurseries, where new stars were forming. For the first time, a more precise estimate was obtained for the age of the universe, about 13.7 billion years. In the process, the HST helped reveal a new mystery.

In the first half of the twentieth century, the telescope's namesake, Edwin Hubble, and other astronomers made a number of remarkable discoveries. One discovery was that the light coming from more distant stars was skewed toward the red end of the spectrum. Called redshift, this is caused by a lengthening or stretching of the emitted electromagnetic wavelengths. What became known as the Hubble-Lemaître Law showed that the further away a galaxy is from Earth, the faster it is receding. Basically, the universe is expanding outward.

Extrapolating backward in time, this expansion suggests that the entire universe, at one time, originated from a single point. The Big Bang became the prevailing theory to describe how the universe rapidly expanded outward, similar to a huge explosion. Early theories held that the expansion of the universe would be decelerating due to gravity. In a similar way, gravity affects long-distance spacecraft like the Pioneers and Voyagers. The probes' speed increased during gravity-assisted slingshot maneuvers around the outer planets, but as they continued traveling outbound through the solar system, the gravitational pull of our massive Sun gradually decelerated them.

72. The space shuttle *Endeavor* orbits high above the Earth in 2010. NASA.

But the HST revealed something very different is happening on the outskirts of the visible universe. The more distant galaxies are not decelerating; they are accelerating. The universe isn't just expanding; it is doing so at an increasing speed. Clearly, some unknown force must exist throughout space, and scientists would spend the following decades trying to unravel this mystery.

Normalcy for NASA

Throughout the 1990s and into the 2000s, the space shuttle continued to service and upgrade the Hubble. Meanwhile, the shuttle program rebounded from the *Challenger* disaster and resumed a sense of normalcy, now operating on a more relaxed schedule, with about seven launches per year. Regular Spacelab missions carried a cylindrical laboratory in the shuttle's cargo bay for conducting microgravity experiments. The Shuttle Radar Topography Mission digitally mapped the planet's elevations from 60° N to 56° S.

Extravehicular Activities, or EVAs, were common. The most dramatic images came from the Manned Maneuvering Unit (MMU), a tetherless rocket chair, used only three times in 1984. After the *Challenger* disaster, the MMU was deemed too risky and reinvented as an emergency jetpack to be worn in case an astronaut broke loose during a tethered EVA. Most trips outside the airlock were focused on recovering, repairing, and assembling orbital equipment.

Other missions were focused on observations toward the stars. More space telescopes were launched, like the Compton Gamma Ray Observatory, the Chandra X-Ray Observatory, and the Spitzer Space Telescope, which focused on infrared observations. Scientists were slowly piecing together a more complete understanding of the universe. The shuttle even released a few interplanetary probes, including *Magellan*, which mapped the surface of Venus. Another was *Galileo*, which observed the collision of Comet Shoemaker-Levy 9 with Jupiter, in 1994, on its way to becoming the first probe to orbit the gas giant. As the decade progressed, the Russian *Mir* space station was completed, and the shuttle began a series of friendly docking missions, often ferrying U.S. and international astronauts to and from orbital stays.

A Russian Proposal

Meanwhile, nearly ten years after authorization, the U.S. space station *Freedom* existed only on paper. New partners had signed on, including the European Space Agency, the Canadian Space Agency, and Japan's NASDA. But there seemed to be little interest from new U.S. president Bill Clinton or Congress in proceeding with the expensive plan.

Then, one day in 1993, NASA received a letter from Yuri Koptev, the general director of the Russian Space Agency, referencing recent meetings discussing the potential for space cooperation. In a halting letter, characteristic of the Russian language's lack of articles, the director mentioned Russia's plans for *Mir-2*, a new-generation space station that would replace the aging *Mir*.

"We are aware about all efforts of USA and its international partners to create space station 'Freedom,'" wrote Koptev. "In this direction there can be indisputable advantages which can be achieved by unification of efforts of Russia and USA. . . . We consider a possibility to suggest for your attention a program of international space station."

The proposal went on to broadly outline a mating of Russian, American, and international modules, high above the earth, to begin a few years before the new millennium. A mix of unmanned Proton and Ariane rockets, manned Soyuz spacecraft, and space shuttle missions could assemble and service this station. Intrigued by the geopolitical possibilities and cost-saving benefits, the Clinton administration agreed.

The first module was launched in 1998, the Russian Zarya, which included propulsion and power systems. Two weeks later, a shuttle delivered NASA's Unity module, attaching this critical docking port during an EVA. In July 2000 the Zvezda module docked automatically, bringing living quarters and life support. This allowed a Soyuz to deliver the first crew later that year.

Further shuttle flights began building the station's unpressurized truss, a linear structure to which large solar arrays could be mounted. Soon came the Pirs docking compartment, the Destiny laboratory, the Quest airlock, and a robotic arm. Early crews were limited to three people, with an emphasis on making the station operational and conducting basic experiments as time allowed.

Columbia Is Lost

By 2003 NASA had flown 112 space shuttle missions, including sixteen visits to the International Space Station (ISS). The partially complete ISS now somewhat resembled a high-tech enlargement of the old *Wright Flyer*, given its rectangular wing-like solar panels. In mid-January, the shuttle *Columbia* launched, with seven astronauts, for a sixteen-day science-focused mission, one of the few that didn't stop at the ISS. During reentry, tragedy struck the shuttle program for the second time.

Passing over Northern California at around 230,000 feet, while traveling over 17,000 mph, *Columbia*'s left wing began to disintegrate. As the returning spacecraft hurtled across the Southwest, observers watched it break apart into streaks of burning debris falling through the sky.

"The *Columbia* is lost," said Pres. George W. Bush, in a televised address to the nation. "There are no survivors."

Once again, the shuttle fleet was grounded. A new investigation revealed disturbing findings. During launch, a piece of spray-on insulation foam had broken off the external tank, striking the leading edge of the orbiter's left wing and damaging the protective thermal tiles. During reentry, hot gases penetrated the wing until it structurally failed, setting off a violent chain reaction that destroyed the entire orbiter. The crew had died from a combination of cabin depressurization, high reentry temperatures, and lethal trauma from sudden movements and flying debris.

73. A camera on board an Apache helicopter captured the orbiter breaking up over Texas. U.S. Army.

Before the accident, NASA managers had spotted the wing strike in launch videos, and some had been concerned about damage to the thermal tiles. But ultimately, these concerns were ignored. The investigation revealed that foam shedding from the external tank was a well-known occurrence at NASA, which dismissed the strikes as an acceptable risk.

The review board report pointed a finger at NASA for developing a culture that disregarded safety, partially based on over-confidence from their past successes. NASA knew the shuttle had design flaws. But over time, when certain flaws like the foam shedding happened without catastrophic results, the organization simply grew to accept these deviations as normal. Using a philosophy called "success-oriented management," NASA was simply hoping for the best.

The report recommended that the space shuttle be retired and replaced as soon as possible. Another part of the Columbia Accident Investigation Board report noted that for more than thirty years, the United States had lacked a guiding vision for space efforts. The lack of a compelling mission, or replacement vehicle for the aging space shuttle, they claimed, "represented a failure of national leadership."

A New Constellation

A year later, President Bush announced a new "Vision for Space Exploration" at NASA headquarters. After safety improvements, the shuttle would return to service to complete construction of the ISS. Then, after thirty years in service, it would be retired. In the interim, a new crew exploration vehicle would be developed to visit the ISS and ultimately travel beyond Earth orbit. America would return to the Moon, establishing an extended presence to launch future missions to Mars and worlds beyond.

During spring 2005, NASA unveiled the Constellation program to meet these objectives. The crew exploration vehicle was called Orion, a conical capsule and service module that the new NASA administrator called "Apollo on steroids." Like the Apollo command module, Orion would launch atop a rocket, reenter the atmosphere using a heat shield, and parachute into the sea for retrieval. But otherwise, this state-of-the-art capsule was much larger, holding up to six crew members, and designed to be reusable. The return of an escape launch system, plus the simpler design and mission profile, meant the vehicle would be much safer than the complex shuttle.

The program would develop two new rockets. Comparable in abilities to the old Saturn IB, the Ares I would include a reusable solid rocket booster (SRB) based on the old shuttle SRBs topped by a liquid-fueled second stage, blending elements of the S-IVB with the shuttle external tank. Nicknamed "The Stick," the Ares I would launch Orion into low Earth orbit. The massive Ares V would be a super heavy-lift cargo rocket, even more powerful than the heralded Saturn V. Carrying an Earth Departure Stage and Altair lunar lander, this cargo vehicle would link up with Orion in Earth orbit before heading to the Moon.

The International Space Station

When shuttle missions tepidly resumed after a two-year hiatus, they came with a variety of safety improvements. Once in orbit, the shuttle thermal tiles would be inspected for damage, with small repairs completed by EVA as

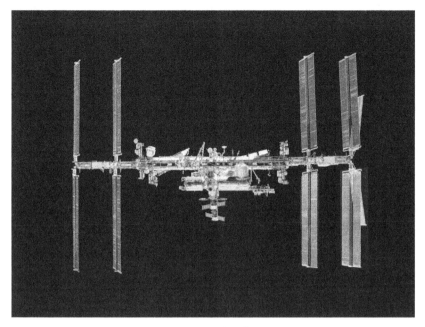

74. This mosaic combines several images taken from the SpaceX Crew Dragon spacecraft upon approach to the International Space Station in 2021. NASA.

necessary. If unrepairable damage was found, contingencies were developed for abandoning ship. All but one mission, to service the Hubble, would be flown to the ISS, where the astronauts could take shelter. A rescue shuttle, kept on standby at the pad, would launch to retrieve them, with Russia offering Soyuz capsules as a backup.

When the return-to-flight mission launched in July 2005, the foam shedding issue happened once again. Fortunately, none of the debris struck the orbiter. The program went into another year-long break for further evaluation. The cause of foam shedding was not human error, as previously determined, but cracks from thermal contraction, during fueling, of below-freezing liquid hydrogen and oxygen. Improvements were made, but the foam shedding issue would remain a concern during the final five years of the shuttle program.

By 2011 twenty-one shuttle missions had concluded America's obligations to the ISS. Though more modules would eventually be added to the ISS via uncrewed launchers, the largest spacecraft ever assembled in orbit was functionally complete: fifteen pressurized modules, a 356-foot truss, and eight massive solar arrays producing over 20 KW of electricity. With an interior volume of 32,000 square feet, the station had space equivalent to that of a 3,500-square-foot building. Weighing almost a million pounds, the ISS orbited 260 miles high and could be seen from Earth with the naked eye.

Continuously crewed since October 2000, the station has been visited by over 250 people from nineteen countries during its first two decades. Crew members typically stayed for about six months, living, working, and regularly exercising to keep their strength in the weightless conditions. Once construction was complete, the focus turned to orbital research. Many studies looked at the long-term effects of microgravity and cosmic radiation on materials, crystals, plants, animals, and the human body. Other observations were pointed toward Earth or out at the stars.

23 Destination Deep Space

From Ion Propulsion to Sample Returns

While the International Space Station (ISS) was developing, around the turn of the twenty-first century, new and exciting uncrewed missions were launching into orbit and spreading out through the solar system. A late-1990s NASA probe called *Deep Space 1* helped test a variety of new equipment, including an ion engine. This highly efficient form of space propulsion uses solar panels to power an electric field that accelerates ions from charged propellant gas as exhaust. The thrust is low (in this case, less than an ounce of equivalent force), and acceleration is slow, but so is the fuel usage. Basically, the ion engine stays on for a long time and slowly gets the spacecraft up to speed. Theorized in the early twentieth century, the first ion thrusters were developed in the United States and Soviet Union during the 1950s and 1960s. The Soviets started using them in the 1970s to stabilize satellites in Earth orbit. *Deep Space 1* was NASA's first major foray, and the engine used only 178 pounds of xenon to increase the probe's speed by 7,900 mph.

About a decade later, NASA would use ion propulsion for an exploratory mission. The *Dawn* probe would visit the two largest objects in the asteroid belt, orbiting both Vesta and the dwarf planet Ceres. *Dawn* ran its engine for just under six years of its eleven-year mission. Its engines used less than 1000 pounds of xenon to change velocity by a total of over 25,000 mph. This record-setting mission proved the deep space abilities of electric propulsion, setting up potential missions to far-flung parts of the solar system.

Other NASA probes included *NEAR Shoemaker*, which orbited and landed on a near-Earth asteroid in 2001. Following up on *Ulysses*, a 1990s solar probe that studied the Sun's north and south poles, the *Genesis* spacecraft collected solar wind particles and returned them to Earth in 2004. NASA's first sample return since Apollo, the *Genesis* return capsule's parachutes failed to open, and it crashed in Utah. Most of the samples were contaminated,

75. This solar-electric ion thruster, in development at NASA's Jet Propulsion Laboratory in Pasadena CA, uses xenon as a propellant. NASA/JPL–Caltech.

but enough material was salvageable to measure the chemical composition of solar wind particles.

A similar mission called *Stardust* looped out to comet Wild 2, in 2004, and collected samples from the comet. This fuzzy envelope of gas and dust surrounds the comet nucleus, giving off the materials that form the tail. After it returned the sample to Earth, the results refined an understanding of how comets form. Another cometary mission, *Deep Impact*, used a mothership to dispatch an impactor that crashed into the surface of Tempel 1 in 2005. Monitoring the crash and crater ejecta helped estimate the density and composition of the Tempel 1 comet.

New Countries in Space

New countries emerged as major space explorers, with the Indian Space Research Organisation using their polar satellite launch vehicle rocket to launch a lunar orbiter, *Chandrayaan 1*, in 2008. An impactor was deployed

76. A solar eclipse, with the earth passing in front of the Sun, as seen during the *Apollo 12* mission in 1969. NASA.

and struck the surface near Shackleton Crater at the lunar south pole. The results helped confirm the presence of water ice on the Moon.

In 2014 the European Space Agency (ESA) used their *Rosetta* probe to deploy the Philae lander onto 67P, thus accomplishing the first soft touchdown on a comet. Unfortunately, upon landing it bounced into the shadow of a cliff. Without solar power, the batteries lasted only two days. But this was just long enough for the probe to detect organic molecules, the building blocks of terrestrial biology. The discovery supported the hypothesis that comet and asteroid impacts may have played a role in the development of life on Earth.

Since their first orbital satellite launch in 1970, the Japanese space program had grown by leaps and bounds. In 2003 the Japan Aerospace Exploration Agency (JAXA) launched their M-V rocket for an ambitious mission. Using ion propulsion, they sent the *Hayabusa* spacecraft to near-Earth asteroid 25143 Itokawa. After landing, the probe collected a small surface sample and returned it to Earth in 2010, revealing its geologic composition. In a sequel, about ten years later, JAXA sent an upgraded *Hayabusa* 2 to asteroid 162173 Ryugu. The mothership deployed three small rovers, which hopped around the low-gravity surface taking temperature readings and photos. The mothership landed and collected another sample, once again returned to Earth in a small capsule.

An American collaboration with the ESA and Italian Space Agency, *Cassini-Huygens* was a celebrated mission to Saturn. After arriving in 2004, the *Cassini* orbiter spent over thirteen years studying the ringed sixth planet and its moons before plunging into Saturn's gaseous atmosphere and burning up. But first, in early 2005, the Huygens lander parachuted down onto Saturn's moon Titan, the furthest landing from Earth ever made. The tiny explorer landed in a dry lake bed or delta. Two photos showed rounded pebbles like those tumbled in a river. Given the temperature was around -290° F, the pebbles could not have been rounded by water but perhaps some other flowing liquid. Ultimately, the mission produced evidence of a methane cycle, existing in frozen, liquid, and gaseous forms. Not only were there methane lakes on Titan, but there were methane clouds, rainstorms, rivers, and seas.

NASA's *Messenger* made a series of complicated maneuvers to reach Mercury, involving flybys of Earth, Venus, and the innermost planet itself, before finally going into orbit in 2011. It was a challenging deceleration given the gravitational pull of the nearby Sun. Just the second spacecraft to visit Mercury, and the first to orbit, *Messenger* solved some mysteries and introduced new ones. Closely resembling Earth's Moon in appearance and size, Mercury is otherwise quite different. The second densest planet in the solar system, after Earth, Mercury has a massive metallic core and relatively thin mantle and crust—unlike any other known planet.

By far, the spacecraft that traveled furthest in the twenty-first century was NASA's *New Horizons*. On its way to the outer solar system, the probe mostly hibernated except for a gravity-assist flyby at Jupiter. When *New Horizons* started its ten-year voyage, its target was the solar system's ninth planet. But during the trip, the International Astronomical Union controversially reclassified Pluto as a dwarf planet. This small orangish sphere, roughly half the size of Earth's Moon, takes 248 years to circle the Sun in a highly elliptical orbit that is mostly further than Neptune but sometimes closer.

Before *New Horizons* became Pluto's first earthly visitor in 2015, early images of the planet were blurry and indistinct. Now up-close photography revealed a strange and diverse world. Most of the surface was nitrogen ice, with mountains of water ice, a smooth heart-shaped plain, and a dark reddish-brown whale-shaped region covered in hydrocarbons. The major planets, dwarf planets, and moons of the solar system had now all been visited, though much remained to be discovered.

Rise of China

Back on Earth, a major new player in space exploration emerged during the early twentieth century, fueled by the explosive economic growth in China. In 2003 the China National Space Administration (CNSA) successfully launched their first taikonaut into space, becoming the third country to independently develop the capability. Carried atop the Long March 2F rocket, the *Shenzhou* spacecraft is based on the Russian *Soyuz*, but enlarged by about 10 percent.

Proceeding at a leisurely pace that allows ample time for testing and development, the CNSA has since rattled off a string of successful crewed and uncrewed endeavors. Missions during the next fifteen years included the first Chinese EVA, long-duration missions in low Earth orbit, and docking with two prototype space stations. Starting in 2021, the CSNA began building the modular *Tiangong* space station and later sent three taikonauts for a ninety-day mission.

As part of the Chinese Lunar Exploration Program, which began launches in 2007, unmanned orbiters, landers, rovers, and sample-return spacecraft have been dispatched to the Moon. Future plans include a crewed lunar mission by the 2030s and eventual establishment of a lunar base. Other aspirations are aimed across the solar system, starting with Mars.

77. The Mars rover *Curiosity* took this composited selfie in front of a rock outcrop called Mont Mercou in 2021. NASA/JPL–Caltech/MSSS.

Return to the Red Planet

Efforts to reach the Red Planet increased throughout the twenty-first century, with missions becoming increasingly international. Unlike the first forty years of exploration, when Mars acquired a reputation as a cursed planet, most recent spacecraft were successful.

NASA's Mars *Odyssey* arrived in 2001 and soon detected large quantities of water ice just beneath the planet's surface. *Mars Express* was the first planetary mission by the ESA, and while the biology-focused lander failed, since 2003, the orbiter has made new discoveries about the planet's atmosphere. NASA's *Mars Reconnaissance Orbiter* arrived in 2006, observed the climate, confirmed further signs of water, and prepared a detailed geologic map.

Arriving in 2004, NASA's *Mars Exploration Rover* mission involved two small solar-powered rovers, each far outlasting its planned ninety-day mission. *Spirit* landed in Gusev Crater and spent the next six years covering almost 5 miles of terrain. Along the way, it revealed evidence of a former lake bed, before getting trapped in loose sand and losing communication in 2010.

Opportunity lasted even longer, exploring 28 miles of Meridiani Planum, which exceeded *Lunokhod 2* and set a record for off-Earth roving. In addition to collecting substantial evidence of past flowing water, the rover stumbled

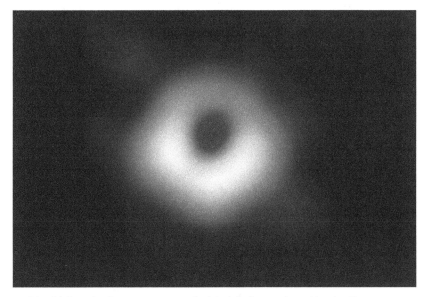

78. This blob is the first-ever image of a black hole, produced by the Event Horizon Telescope. EHT collaboration.

across a basketball-sized meteorite, the first found on another planet. *Opportunity* hibernated during a planet-wide dust storm in 2018 but never woke up, which brought its remarkable fifteen-year mission to an end.

In 2012 NASA returned to the surface with the Mars Science Laboratory, which landed the celebrated *Curiosity* rover inside massive Gale Crater. Powered by an RTG (radioisotope thermoelectric generator), this car-sized rover weighs almost 2,000 pounds and far outlived its two-year mission, remaining operational to this day. During its explorations, *Curiosity* traveled 17 miles, returning stunning images and video. Ground studies revealed the presence of organic chemicals and evidence of an ancient mega-flood that ripped across the landscape billions of years ago.

More missions followed, like NASA's *Maven*, which studied upper atmosphere gases. The Indian Space Research Organisation became the fourth space agency to arrive with the Mars Orbiter Mission in 2014, the first interplanetary mission for India, which succeeded on its first attempt. The ESA and Roscosmos (renamed and restructured during the presidency of Vladmir Putin), collaborated on the ExoMars program in 2016. One part, the Trace Gas

Orbiter, studied atmospheric methane, possibly created by past biological activity, but the Schiaparelli lander crashed into the surface. In 2018 NASA's InSight lander placed a seismometer on the surface, which documented active marsquakes and revealed a molten core.

A probe from the United Arab Emirates, with the help of a Japanese launcher, reached Mars orbit in 2021 to study its weather. China's first mission also arrived in 2021, with the *Tianwen-1* orbiter and the *Zhurong* rover, It was the third country to soft land on the planet's dusty surface. That same busy year, NASA landed the *Perseverance* rover, similar to *Curiosity*, with a new twist called Ingenuity. This small solar-powered helicopter, which carries a commemorative scrap of wing material from the 1903 *Wright Flyer*, became the first robotic aircraft to conduct powered flight above another world. Future missions are ongoing, making Mars the preferred destination beyond the Earth-Moon system, helping pave the way to future human visitation.

24 The Billionaires' Space Race

The NASA Gap

While uncrewed missions proliferated throughout the early twenty-first century, and new space agencies made increasingly bold strides, crewed spaceflight at NASA mostly languished. During the final few years of the space shuttle program, the follow-up Constellation program was increasingly criticized for being behind schedule, over budget, and underfunded, having never received enough support to match its goals. After five years, the Ares I crew launch vehicle had seen only one uncrewed test flight, mostly successful but suffering severe pogo oscillation that might have killed human occupants. Meanwhile, the Ares V rocket existed mostly on paper, and the *Orion* spacecraft was taking longer than expected.

New president Barack Obama appointed a panel to review the program, which concluded that most of Constellation's goals were unfeasible. The panel recommended reassessing U.S. space priorities to better match budget realities, including reducing the scope of Constellation by perhaps focusing on either a Moon landing or a Mars landing, but not both. Other considerations were searching for off-world resources and increasingly relying upon the burgeoning private spaceflight industry.

After reviewing the report, the Obama administration caused an uproar by removing the entire Constellation program from the 2011 budget. This effectively canceled the Ares rockets and *Orion* spacecraft, leaving NASA to rely entirely on private firms to develop new vehicles to reach LEO and the International Space Station (ISS). Following widespread criticism, particularly around the lack of any U.S. space exploration goals, President Obama visited Kennedy Space Center in 2010 to outline a new program.

The plan involved continuing the development of *Orion*, bypassing a return to the Moon, and focusing on sending astronauts deeper into space—first to an asteroid and later Mars. To do this, a single super heavy-lift rocket would

79. The space shuttle *Endeavor* lands at night at Kennedy Space Center after the STS-72 mission in 1996. NASA.

be developed. Members of Congress, upset over the loss of constituent jobs in the Constellation program, got involved by mandating that the new rocket prioritize existing space shuttle and Constellation contracts.

The result was the Space Launch System (SLS), the first U.S. rocket created by an act of Congress. In many ways a modification of the Ares V, the SLS has a core stage based upon the space shuttle's external tank, with four RS-25D engines previously flown on the orbiter. Two expendable solid-fuel boosters were based on the shuttle SRBs. The *Orion* spacecraft was to be redesigned so that one multipurpose version could accommodate the various missions, and service module development was assumed by the European Space Agency.

In 2011, after thirty years of service, the space shuttle program was retired. Five shuttles had flown a total of 135 missions, with 355 astronauts from sixteen countries spending 1,323 days in orbit. But two orbiters had been lost, along with fourteen lives, in the *Challenger* and *Columbia* disasters. Without a way to reach or return from orbit, U.S. astronauts would now rely on paid seats on Russian Soyuz flights, while NASA worked with private companies developing new launch vehicles. Fortunately, the private U.S. spaceflight industry had made major strides in recent years.

From Ansari X Prize to Privatization

Back in 1996, the entrepreneurial Ansari family created the X Prize, a $10 million award to the first private company to launch a reusable crewed spacecraft into space twice within two weeks. Twenty-six teams from across the world participated in the competition. Eight years later, engineer Burt Rutan won the prize with his experimental spaceplane, *SpaceShipOne*.

With the fuselage sharing a bulletlike resemblance to the original Bell X-1, the SS1 is carried aloft by a launch aircraft. At an altitude of around 45,000 feet, the SS1 detaches and fires its rocket engine at a 65-degree angle, which carries it at 2,685 mph to an altitude of over 60 miles high. During the apex of its ballistic trajectory, the space plane experiences several minutes of freefall.

Its boxy wings, with horizontal and vertical surfaces, rotate upward into a high-drag mode, which helps slow the aircraft during its fall through the atmosphere. Nearing the surface, the wings return to horizontal glide mode for landing. During fall 2004, *SpaceShipOne* made two flights five days apart, winning the Ansari X Prize.

Six years later, the company Scaled Composites began testing *SpaceShipTwo*, an improved version twice the size of the original. The goal was for the new spaceplane to be flown on commercial space tourism flights operated by Virgin Galactic. Early tests were unpowered, with the first rocket-powered flight happening in 2013. During a test flight the following year, the wings were errantly switched into feathered reentry mode prematurely, causing the spaceplane to break apart. The injured pilot parachuted to safety, while the copilot was killed—the first crash and fatality in crewed commercial spaceflight.

A replacement vehicle was built, making its first crewed flight from New Mexico's Spaceport America in July 2021. Among the passengers was company owner Sir Richard Branson. His flight upstaged, by just eight days, the first flight of another private space company owner, Jeff Bezos. Branson and Bezos, along with Elon Musk of SpaceX, were engaged in a new competition: the so-called Billionaire Space Race.

In 2000 the owner of Amazon, Jeff Bezos, founded the private aerospace company Blue Origin. Working mostly in secret, the company began developing rocket engines and launch vehicles, initially for the purpose of suborbital space tourism to fund later developments like orbital satellite launches. In

80. The authentic *SpaceShipOne*, which won the Ansari X Prize in 2004, on display at the Smithsonian Air and Space Museum in Washington DC (2021). Mike Bezemek.

2015 the company revealed their New Shepherd rocket, which made a series of successful test launches over the coming years. This paved the way for Bezos and three passengers to launch on a suborbital flight in July 2021, just eight days after Branson's flight. Accompanying Bezos was eighty-two-year-old Wally Funk, one of the Mercury 13, a group of female pilots who were part of a privately funded astronaut screening test program in the 1960s. A later suborbital launch carried ninety-year-old William Shatner, the famous actor behind *Star Trek*'s Captain Kirk.

While Blue Origin and Virgin Galactic were making progress, during the 2010s, toward establishing a lucrative suborbital tourism business, NASA was still lacking a reliable way to reach the ISS. Fortunately, another company, with ever greater ambitions, had emerged as the leader in private spaceflight.

SpaceX & Vertical Landing

The Space Exploration Technologies Corporation was founded in 2002 by former PayPal CEO Elon Musk. Called SpaceX, the company's goal was to

81. A Falcon 9 booster stage returns for a vertical landing in 2022. SpaceX.

reduce the cost of space transportation by pioneering reusable launch systems that would ultimately enable the colonization of Mars. The first vehicle was an expendable orbital launch rocket, the Falcon 1. At 69 feet tall, with a single engine producing about 80,000 pounds of thrust, in some ways it was like a modern version of von Braun's Redstone. In this case, the lone Merlin engine burned liquid oxygen and RP-1, a refined form of kerosene that is considered the standard in rocket fuel. The Falcon 1 had a second stage designed to push a small 400-pound payload into low Earth orbit (LEO).

Launched from Kwajalein Atoll in the Marshall Islands, the first test flight, in 2006, partly exploded. The next year a redesigned second rocket shut down early and failed to reach orbit. A third attempt, in 2008, ended in a similar failure. With the company running low on money, they pushed ahead with a fourth test launch only six weeks later, possibly the last, should it also fail. This time the Falcon 1 succeeded, becoming the first privately funded liquid-fueled rocket to reach orbit. After a fifth successful flight with its first paying customer, marking the first private satellite launch, the company was saved when NASA offered a $1.6 billion contract to make twelve future cargo flights to the ISS. Recognizing the limitations of the Falcon 1, SpaceX proceeded to developing a next-generation rocket.

The Falcon 9 stood 180 feet tall, weighing over ten times more than its predecessor. Its nine first-stage Merlin engines produced 1.1 million pounds of thrust, giving its second stage the capacity to deliver 23,000 pounds to LEO—slightly more than the Saturn I. Now launching from Cape Canaveral, the first test flight was a success. A few months later, a second flight sent up the new Dragon cargo capsule, a partially reusable spacecraft that used a replaceable heat shield. After testing maneuvering thrusters in orbit, the Cargo Dragon made a successful reentry and splashdown. In May 2012 a third Falcon 9 successfully sent a Cargo Dragon to dock with the ISS, completing the first commercial supply mission. After two more visits to the ISS, SpaceX introduced an upgraded Falcon 9 v1.1, about 44 feet taller with an extra 6,000 pounds of payload capacity, and a few revolutionary additions.

During this sixth flight, after the second stage carried the Falcon 9's first commercial payload to orbit, the core stage performed a controlled reentry. Basically, thrusters oriented the rocket nozzles toward the ground, and three engines were reignited to decelerate the freefall. Instead of burning up, the maneuver brought the rocket down through the atmosphere in one piece. As it neared sea level, it started to roll, the engines shut down, and it made a hard impact into the ocean.

A few flights later, using data from the previous test and updated thrusters, they improved the maneuver in 2014. This time, the core stage came down vertically, and as it approached the ocean's surface, four landing legs deployed. Several months later, they tried again. This time, they used the reignited engines to boost the core stage on a reverse trajectory, soft-landing it in the sea not far from Cape Canaveral. In early 2015 a Falcon 9 performed another controlled reentry, this time trying to land the core stage on an autonomous floating platform at sea. Upon approach, it lost control and exploded on impact. More drone ship failures would follow, often with the core stage coming close to landing, but tipping over and erupting in a fireball.

Called vertical takeoff and vertical landing, the concept had been discussed for decades and had been repeatedly imagined in early science fiction. The first attempt to implement the approach was an experimental triangular rocket called the DC-X. In the mid-1990s, a one-third scale prototype flew a dozen times from White Sands Space Harbor in New Mexico. During the course of testing, the DC-X rose thousands of feet high, hovered, and became the first rocket to land vertically. But during the twelfth test, a landing strut failed to deploy. The vehicle tipped over and exploded, effectively ending the project.

In November 2015 the secretive company Blue Origin surprised observers by releasing a video of their New Shepard executing the first vertical landing of a suborbital rocket. In announcing the feat, founder Jeff Bezos made a subtle dig at upstaged rival Elon Musk, who retorted about the easier nature of slower-moving suborbital returns. A month later, SpaceX achieved an even more challenging feat—the first vertical landing of a much faster moving orbital rocket onto a landing pad at Cape Canaveral. The next year, the company finally nailed a drone ship landing at sea.

In 2017 SpaceX launched the first previously flown core stage. Having developed partially reusable orbital launch technology, SpaceX's cost to LEO would soon plummet. When accounting for inflation and using 2022 dollars, the space shuttle's 134 launches had averaged to about $30,000 per pound to low Earth orbit. Since 1997 the Ariane 5G's twenty-four launches have averaged about $4,600 per pound. Since 2010 the Falcon 9 has averaged less than $1,200. By 2018 SpaceX was launching more rockets than any other single entity in the world, including private companies and governments.

SpaceX unveiled the Falcon Heavy in 2018. This super heavy-lift launch vehicle combined three modified Falcon 9 core stages to produce 3.4 million pounds of thrust. During a successful test flight, the upper stage escaped Earth's gravity and entered a heliocentric orbit carrying a surprising dummy payload: Elon Musk's Roadster sports car, manufactured by his electric car company, Tesla Motors. Strapped into the driver's seat was a mannequin, nicknamed Starman, wearing a SpaceX pressurized spacesuit. In a publicity coup, photos of the driver and red sports car, with the top down, floating above the earth, were viewed all around the world.

In 2020 SpaceX achieved a milestone previously accomplished by only three nations, the U.S.S.R. or Russia, the United States, and China. The SpaceX Crew Dragon capsule launched atop a human-rated Falcon 9 with two NASA astronauts aboard. Upon docking at the ISS, it was the first crewed visit from the United States since the final space shuttle mission nine years before. After two months aboard the station, the astronauts safely splashed down in the

82. A prototype SpaceX Starship launches on SN-15, a successful high-altitude test flight, in 2021. SpaceX.

Gulf of Mexico. Regular Crew Dragon flights followed, ferrying astronauts to and from the ISS. Meanwhile, a three-day orbital flight carried four private citizens as a charity effort for St. Jude's Children's Hospital.

The Future of Space Travel

Throughout the early 2020s, much attention turned to the development of the SpaceX *Starship*. If completed as designed, this fully reusable two-stage super heavy-lift launch vehicle will become the biggest rocket ever built. Projected to be about 393 feet tall, it will exceed the Saturn V and planned Space Launch System Block 2 Cargo rocket.

Mounted atop the stainless-steel booster will be the *Starship* spacecraft. This bullet-shaped vehicle will be capable of refueling in orbit and carrying people and cargo to the Moon and onward to explore and potentially colonize Mars. Using new Raptor engines, *Starship* burns liquid oxygen and liquid methane, chosen because both propellants could be made from resources available on the Red Planet.

Using four control flaps and a tiled heat shield, *Starship* will return to Earth by falling through the atmosphere, performing a belly flop maneuver, and firing its engines in retro-orientation to land vertically. The first launch to Earth orbit is hoped for during 2023 or 2024, with an anticipated uncrewed flight to Mars coming some number of years later, and a crewed flight after an unknown interval.

Another development in U.S. space efforts came in 2017, when Pres. Donald Trump reinstituted several programs canceled by his predecessor. The result was the Artemis program, which would make use of the SLS and *Orion* spacecraft to return Americans to the Moon during the 2020s, including landing a woman and a person of color. Uncrewed and crewed test flights around the Moon would be followed by a crewed lunar landing and a visit to the *Lunar Gateway*, a small space station placed in orbit around the Moon. The mission would use a mix of NASA and SpaceX hardware, with a version of *Starship* serving as a lunar lander.

Elsewhere in near-Earth space, a new infrared space observatory was launched in late 2021. After traveling about 930,000 miles beyond Earth's orbit around the Sun, the James Webb Space Telescope unfolded while orbiting Lagrange Point 2. Like the Hubble, or the collaborative Event Horizon Telescope, which produced the first image of a black hole in 2019, the Webb soon was returning stunning images of the cosmos that will lead to new discoveries about the universe. Perhaps its results will continue to reignite a growing interest in space.

Meanwhile, other missions are aiming at the Moon, asteroids, and planets. Mercury, Venus, Mars, and Jupiter will all see new missions. The *Europa Clipper* will study the icy Jovian moon for signs of life, while *Dragonfly* will place a quadcopter drone on Saturn's moon Titan. Though nothing is yet planned, even the ice giants, Uranus and Neptune, will eventually be visited by orbiters.

Midway through the third decade of the third millennium, longtime space enthusiasts look forward, and upward, to an exciting era in space exploration. At the same time, new enthusiasts are made daily, inspired by recent events like the return of U.S. space launches and the coming of new possibilities, like returning to the Moon and sending people to Mars. For these reasons, there's never been a better time to look back at the remarkable achievements of the space age, to explore the fascinating artifacts at the many museums and historic sites, and to learn about the dramatic stories behind them.

It has been almost one hundred years since Robert Goddard launched the first liquid-fueled rocket from a farm field in Massachusetts. Roughly eighty years since von Braun's V-2 pierced the edge of space on its way to terrorizing Allied cities like London. About sixty-five years since Korolev's R-7 rocket carried Sputnik into orbit, followed by Yuri Gagarin five years later. Roughly fifty years since von Braun's Saturn V burst into the sky, carrying Neil Armstrong to that first small step. Just over forty years since the space shuttle first launched, and more than ten since the surviving orbiters were retired to museums. In just the last decade, space travel has shifted from being solely the endeavor of national government agencies to including the personal ambitions of billionaire-owned private companies.

As these early eras of exploration have shown, humanity's thrilling future among the stars is only beginning, and countless space age adventures await, just beyond the horizon.

83. The ESA Solar Orbiter launches from Cape Canaveral on a mission to study the Sun in 2020. NASA/Jared Frankle.

Sites on the West Coast
Nineteen Adventures and Seven Orbital Neighbors

Celestial Overview: The West Coast, and California in particular, has a storied history associated with aviation and space travel. For that reason, this region accounts for the second-highest number of space trips in this book. The state is home to two of the ten NASA research facilities found around the country, the Ames Research Center and the Jet Propulsion Laboratory, the latter of which can be visited on a guided tour arranged in advance.

Elsewhere in Southern California, a must-see destination for space enthusiasts is the California Science Center in Los Angeles. In addition to having a complete authentic shuttle stack, the center has a space-flown capsule from each of NASA's first three human spaceflight programs. When you're in town, check out the nearby Griffith Observatory and consider stopping at smaller museums and science fiction film sites. Continuing north along the West Coast leads to many additional space highlights.

San Diego Air and Space Museum, San Diego CA

Debriefing: Located in Balboa Park, the San Diego Air and Space Museum is a large facility with an impressive exhibition of aircraft, about a half dozen worthy spacecraft, and other space artifacts and exhibits.

Most exhibits present various eras in aviation, including early milestones, the Great War, the golden age of flight, World War II, and the modern era. The museum's aviation highlights include several Wright brothers planes and a flying replica of the *Spirit of St. Louis*, the monoplane flown by Charles Lindbergh for the first nonstop flight from New York to Paris. There's also an F-18 Blue Angel, a Predator drone, and a mock-up of the Bell X-1, the rocket-powered aircraft that was first to intentionally break the sound barrier in controlled, level flight.

The STARS: Highlights include the space-flown *Apollo 9* command module, named Gumdrop, which was used to qualify the lunar module for lunar orbit operations prior to the first Moon landing.

Hanging in the Modern Jet and Space Age Gallery, there's a full-scale model of the Apollo command and service modules. Nearby, there's an Apollo lunar sample container and a Moon rock from *Apollo 17*. Other full-size mock-ups include the complete Mercury and Gemini spacecrafts. Rounding out the displays are several spacesuits and a Boeing GPS-12 satellite.

EVAs: While in the area, make sure to check out the rest of Balboa Park, which has great paths for walking and cycling and several other excellent museums, and the San Diego Zoo.

Countdown: Open daily 10:00 a.m.–4:30 p.m.

Mission budget: $$

Flight team: 619-234-8291 | https://sandiegoairandspace.org

Coordinates: 32.7262, -117.1544 | 2001 Pan American Plaza, San Diego CA, 92101

Flight plan: The museum is in the southwestern corner of Balboa Park, and free parking is available on-site.

Rations: The **Flight Path Grill** is the museum's on-site restaurant, typically open on weekends. There are several other restaurants inside the park, with plenty more throughout the city.

California Science Center, Los Angeles CA

Debriefing: The California Science Center combines permanent exhibits on topics like life science and ecosystems with a small aviation offering and an impressive exhibition of space-flown artifacts. On the aviation front, there's an A-12 Blackbird, a T-38 Talon, a F/A-18 Hornet, a full-scale flying replica of a 1902 *Wright Glider* and more. The center also shows IMAX films, typically with at least one related to space, and hosts a series of special exhibitions, so check the website for current offerings.

The STARS: The center possesses authentic examples of all three major components of the space shuttle stack, including the *Endeavour* orbiter, the final flightworthy external tank, ET-94, still in existence, and a pair of previously flown solid rocket boosters (not currently on display). The center's long-term plans are to construct a new 200,000-square-foot exhibit hall, the Samuel Oschin Air and Space Center, where they will display all space-related exhibits and artifacts, including the full shuttle stack in its vertical launch configuration. Until then, the space shuttle *Endeavour* is being displayed horizontally, while the external tank is on display outside the center.

A related exhibit, *Mission 26: The Big Endeavour*, features photos of transporting the orbiter to the center, including its flight over California and a twelve-mile trip on city streets through urban Los Angeles.

Other artifacts include the flown Mercury-Redstone 2 capsule, which carried the chimpanzee Ham on a suborbital flight, to qualify the spacecraft for human space travel, prior to Alan Shepard's first flight.

Nearby is the flown *Gemini 11* capsule, which carried Dick Gordon and Pete Conrad into orbit for three days in 1966, where they set records for altitude (850 miles).

And there's the Apollo-Soyuz command module, flown in 1975 by Tom Stafford, Deke Slayton, and Vance Brand to an orbital rendezvous with a Soviet Soyuz spacecraft carrying Alexei Leonov and Valery Kubasov—the first international human space flight mission.

Rounding out the space-related exhibition is a series of models. A full-scale engineering model of the Viking lander, which made history as the first successful spacecraft landing on Mars, in 1976. Full-scale engineering models of *Cassini*, the first spacecraft to orbit Saturn, and its *Huygens* probe, which landed on the surface of Titan. There's also a one-fifth-scale model of the Chandra Space Telescope and a one-fifth-scale model of the Hubble Space Telescope.

EVAs: The center offers a variety of simulators on rotating topics related to exhibits, including one about the space shuttle program called the Endeavor Together Simulator. Summer camp classes related to space may be available— call the center for more information.

Countdown: Open daily 10:00 a.m.–5:00 p.m.; closed major holidays

Mission budget: $–$$ (admission is free but parking and special exhibits require tickets)

Flight team: 323-724-3623 | https://californiasciencecenter.org/

Coordinates: 34.0158, -118.2863 | 700 Exposition Park Drive, Los Angeles CA, 90037

Flight plan: The **California Science Center** is in Exposition Park, and a paid parking facility is available. Public transportation options include a Metro stop outside the museum.

Mission parameters: The center may be very busy during weekends and during weekday mornings, due to field trips—weekday afternoons are often

84. On its way to the California Science Center, the space shuttle *Endeavor* is flown atop a NASA 747 over Griffith Observatory. NASA/Jim Ross.

quieter. The 160-acre **Exposition Park** offers more museums and attractions to visit, including the **California African American Museum,** the **Natural History Museum of Los Angeles County**, and the **City of Los Angeles Rose Garden**.

Rations: The center offers three on-site eateries, the **Trimana Grill, Market, and Coffee Bar**. Additional restaurants abound throughout the area.

Griffith Observatory, Los Angeles CA

Debriefing: Located above Los Angeles, at an elevation of 1,134 feet, on Mount Hollywood in Griffith Park, this free facility is a public astronomy observatory, planetarium, and museum. Griffith Observatory opened in 1935 and was renovated in 2006 and is operated by the city of Los Angeles.

The STARS: In the Keck Central Rotunda is the large Foucault Pendulum and eight restored wall murals, created by Hugo Ballin in 1935, that celebrate celestial mythology and the advancement of science. The Wilder Hall of the Eye presents four exhibit areas related to observing the sky by naked eye, telescopes, observatories, and spacecraft.

The Ahmanson Hall of the Sky explores the stellar relationship among the Sun, Moon, and Earth, including thermal radiation, the seasons, tidal forces, and more. The Cosmic Connection presents a timeline of the history of the universe from the Big Bang to the formation of the Milky Way and the evolution of life on Earth.

The Edge of Space includes samples of meteorites and models of asteroids, a large Moon globe, and an interactive telescope exhibit. The Gunther Depths of Space exhibit includes informational displays and models of Earth and other planets in the solar system.

The Samuel Oschin Planetarium is a 290-seat domed theater offering a variety of hosted star shows every sixty to ninety minutes when the observatory is open. The planetarium uses the renowned ZEISS Universarium Mark IX, one of only three such optical star projectors in the United States. (The other two are at the St. Louis Science Museum and the Hayden Planetarium in New York.) A show schedule is available on the observatory website, but tickets can only be purchased at the observatory on the day of the show.

EVAs: Public telescope viewing is available every night the observatory is open, provided skies are clear, typically beginning about 7:00 p.m.

Countdown: Friday 12:00–10:00 p.m.; Saturday and Sunday 10:00 a.m.–10:00 p.m.; closed Monday–Thursday

Mission budget: Free for exhibits ($ for planetarium)

Flight team: 213-473-0800 | https://griffithobservatory.org

Coordinates: 34.1186, -118.3004 | 2800 East Observatory Road, Los Angeles CA, 90027

Flight plan: The observatory is located atop Griffith Park, with paid parking on-site. Public bus transportation is available.

Rations: The **Café at the End of the Universe** is the observatory's on-site restaurant and lunch counter, open until 6:00 p.m.

More Space Sites, Southern California

Debriefing: There are several additional sites throughout Southern California with space-related significance. At the top of the list is a tour of a famous NASA research center, followed by a NASA deep space communications complex, a pair of astronomical observatories, several small museums, two sites at Edwards Air Force Base, and one air and space port. Some of these sites require advance reservations or are open only on certain days.

The STARS: Located in Pasadena, NASA's **Jet Propulsion Laboratory** is a research and engineering facility focused on producing robotic spacecraft and conducting other space missions. JPL offers a limited number of free tours for groups and individuals by reservation only. Reservations must be made at least three weeks in advance. **Flight team:** https://www.jpl.nasa.gov/events/tours

Located 5,715 feet high in the San Gabriel Mountains near Pasadena, the **Mount Wilson Observatory** is an astronomy facility operated by the Carnegie Institution for Science. The observatory has a historic 100-inch telescope, a 60-inch telescope, two solar telescopes, and more. There's a small astronomy museum, and public tours are available at 11:30 a.m. and 1:00 p.m. on weekends. **Flight team:** https://www.mtwilson.edu/

Located in Downey, a suburb of Los Angeles, the **Columbia Memorial**

Space Center is a small interactive aerospace museum dedicated to the space shuttle *Columbia*, with exhibits about space and science. **Flight team:** www .columbiaspacescience.org

Located on Palomar Mountain, northeast of San Diego, the **Palomar Observatory** is an astronomic research center owned and operated by Caltech. Its three research telescopes include a 200-inch, 60-inch, and 48-inch. The visitor center includes a small astronomy museum, and guided tours are available. **Flight team:** https://sites.astro.caltech.edu/palomar

Located in Barstow in the Mojave Desert, the NASA **Goldstone Deep Space Communications Complex** handles two-way communication with space probes like *Voyager 1*, the *Mars Reconnaissance Orbiter*, and *New Horizons*. Goldstone is one of three such facilities, with the other two located near Madrid, Spain, and Canberra, Australia. The visitor center and museum are open to the public Mondays, Wednesdays, and Fridays 9:00 a.m.–3:00 p.m. A limited number of guided tours are available by advance reservation. **Flight team:** https://www.gdscc.nasa.gov/

Located in the Mojave Desert, Edwards Air Force Base includes the **Air Force Flight Test Center Museum** and Armstrong Flight Research Center.

85. Astronomer Carl Sagan stands with a replica Viking Mars lander in Death Valley. NASA/JPL.

Currently, the only way to visit these sites is during a public tour—though the air force is in the process of building an impressive public museum at a site just outside the base. In the meantime, community engagement tours were canceled during the COVID-19 pandemic but may resume. **Flight team:** www .edwards.af.mil/Contact-Us/Community-Engagement/Tours/

Located in the city of Mojave, the **Mojave Air and Space Port** is used as a testing facility by a variety of private space companies. The port is open to the general public only on the third Saturday of every month, called Plane Crazy Saturdays. During the free event, visitors can view historic aircraft, the Roton rotary rocket, and replicas of *SpaceShipOne* and the *Rutan Voyager*, the first plane to fly around the earth without stopping or refueling. **Flight team:** http://mojavemuseum.org

Located in the Mojave Desert town of Boron, the **Saxon Aerospace Museum** is a small volunteer museum, focused on flight milestones, in the Edwards Air Force Base area. These milestones include the Bell-X1 breaking the sound barrier, the first hypersonic flight, and the first space shuttle landing. The museum shut down due to COVID-19 but may reopen in the future. **Flight team:** www.saxonaerospacemuseum.com/

Orbital neighbor: Located on the campus of the University of Hawaii at Hilo, on the Big Island of Hawaii, the ʻ**Imiloa Astronomy Center** is the educational outreach facility for the telescope observatories at the summit of Mauna Kea. Exhibits cover Mauna Kea, astronomy, the solar system, and Hawaiian culture and history. A planetarium and 3D dome theater offer star shows and films. Tours to the Mauna Kea summit were suspended due to COVID-19 but may resume in the future. Contact the center for more information. **Flight team:** https://imiloahawaii.org

Science Fiction Film Sites, California

Debriefing: With much of the motion picture industry located in Hollywood, many impressive film sites can be found throughout Southern California, with some extending into the northern parts of the state. Here are some production locations that may appeal to space enthusiasts.

The STARS: Located in the Sierra Pelona Mountains north of Los Angeles, **Vasquez Rocks Natural Area** is an LA County park made famous as an outdoor film set in hundreds of movies and TV shows. In particular, science fiction fans may recognize the tilted sedimentary rock formations from the late 1960s series *Star Trek*. Nicknamed Kirk's Rock, the formation was the backdrop when William Shatner's iconic Captain Kirk battled a reptilian humanoid from a species called the Gorn. Vasquez Rocks was subsequently used in many other *Star Trek* films and TV series. Other productions include *The Twilight Zone*, *Battlestar Galactica*, and *Galaxy Quest*. Today the park remains a frequently used film set, while a series of trails allow visitors the chance to explore the site and scramble up the rocks. **Flight team:** https://parks.lacounty.gov/vasquez-rocks-natural-area-and-nature-center/#

Located on the east side of the Sierra Nevada Mountains, near the town of Lone Pine, the **Alabama Hills** are a formation of rounded granite rocks used as a film location for hundreds of movies and TV shows. Most of them are Westerns, but the Alabama Hills have frequently appeared in various *Star Trek* films and TV episodes, plus other sci-fi productions. Today the area is popular for hiking, camping, visiting natural stone arches, and exploring the areas called Movie Flat and Movie Road. **Flight team:** https://www.blm.gov/visit/alabama-hills

Located in southeastern California, **Death Valley National Park** is not only a great place to explore a stunning desert landscape, but it was once a popular film location, most notably in 1977's *Star Wars: A New Hope* and the 1983 sequel, *Return of the Jedi*. While most of the Tatooine scenes were shot in Tunisia, other scenes involving R2-D2 and C3PO were shot around Death Valley. Filming sites include the Mesquite Flat Sand Dunes, Artists Palette, Dantes View, Desolation Canyon, Golden Canyon, and Twenty-Mule Team Canyon. Other productions at Death Valley include *The Twilight Zone* and many Westerns. With a shift to preserving much of the park as wilderness, film production has mostly ceased at Death Valley. Today, visitors can drive scenic roads, visit viewpoints, and explore the park and film sites on hiking trails. **Flight team:** https://www.nps.gov/deva/index.htm

Other sites around California include Fern Dell Nature Trail, located at Griffith Park in Los Angeles, which was featured in episodes of *Star Trek: The Next Generation* and *Deep Space Nine*. Nearby, the Griffith Observatory was a setting in the original 1986 *The Terminator*, plus many other films. Point Dume State Beach, in Malibu, was the bluff-lined beach in the final scene in *Planet of the Apes* with Charlton Heston.

In Northern California, Yosemite National Park made an appearance in *Star Trek V: The Final Frontier*. In Grizzly Creek Redwoods State Park, Cheatham Grove was used as the forest moon of Endor in *Star Wars: Return of the Jedi*. Nearby, scenes of *E.T. the Extra Terrestrial* were shot in and around Redwood National and State Parks. And Fern Canyon in Prairie Creek Redwoods State Park (permit required to visit during high season) was a film location for *Jurassic Park: The Lost World*.

Chabot Space and Science Center, Oakland CA

Debriefing: Located in Redwood Regional Park in Oakland, the Chabot Space & Science Center is an interactive facility with hands-on exhibits, a small display of space artifacts, a large-screen theater, and a planetarium. Originally opened in 1883 as the downtown Oakland Observatory, the astronomy facility moved in 1915 to its present location at an elevation of 1,500 feet to avoid light pollution. Since that time, Chabot has evolved into an interactive educational center, hosting regular programs, with three large telescopes available for public use.

The STARS: The NASA Experience is an interactive exhibition related to efforts at the Ames Research Center, involving hands-on challenges like building rovers and robots. Part of the exhibit includes thirty space objects, like spacesuits from Projects Mercury and Gemini, experimental heat shield technology, a wind tunnel fan blade, and more.

Going the Distance is an interactive exhibit tracing humanity's past, present, and future journeys through space. A model mission control allows participants to pilot a rover across a Mars-like planet. Artifacts include some small satellites and an actual Russian Soyuz descent module.

Touch the Sun is an interactive animation display of the Sun, allowing visitors to zoom in on solar features, including hot spots, prominences, and possibly flares or coronal mass ejections.

A 70-foot domed planetarium offers a variety of shows starting each hour when the center is open. Shows range in length and topic and are included with general admission.

EVAs: The Chabot Observatories are the center's three large telescopes located on-site. The Leah telescope is an 8-inch refractor built in 1883. The 1915 Rachel telescope is the largest refractor in the western U.S. that's regularly open to the public. The newest and most powerful is the Nellie telescope, a 36-inch reflector that offers access to 180 degrees of the night sky. Telescope viewings are free and open to the public every Friday and Saturday night, weather permitting, from 7:30 p.m. to 10:30 p.m.

Countdown: Saturday and Sunday 10:00 a.m.–5:00 p.m.

Mission budget: $$

Flight team: 510-336-7300 | https://chabotspace.org

Coordinates: 37.8189, -122.1807 | 10000 Skyline Boulevard, Oakland CA, 94619

Flight plan: Located in Redwood Regional Park, about five miles east of downtown Oakland, the center has free parking on-site.

Rations: There's no food at the center, but countless options can be found throughout the East Bay.

Orbital neighbor: Outside of San Jose CA, the **Lick Observatory** is a telescope research facility affiliated with the University of California. Located atop Mt. Hamilton, there's a daytime visitor center, exhibits, and outside walking tours. Nighttime telescope viewing is available by purchasing tickets in advance. **Flight team:** www.lickobservatory.org

USS *Hornet* Sea, Air, and Space Museum, Alameda CA

Debriefing: The focus of this museum is on the USS *Hornet*, an Essex-class aircraft carrier in service between 1943 and 1970. The *Hornet* participated in operations during World War II and the Vietnam War. During the *Apollo 11* and *12* missions, the Hornet was the primary recovery vessel, with the astronauts and capsule collected in the Pacific Ocean for transport to Hawaii. Today, the aircraft carrier has been converted into a floating museum, docked in Alameda, on the eastern side of San Francisco Bay. In addition to touring the aircraft carrier, there are fifteen aircraft and several interesting artifacts from the Apollo program.

The STARS: The main space highlight is the mobile quarantine facility, originally built as a prototype model. It was updated to quarantine the crew of *Apollo 14*.

Nearby, there's a boilerplate test capsule from Project Gemini, and an Apollo test capsule, which was launched for a single orbit in 1966 and recovered by the USS *Hornet*.

Other items include a small lunar module concept model called prototype D and a Sikorsky Sea King Helicopter, which recovered the crew of *Gemini 4* in the Atlantic Ocean in 1965.

EVAs: The museum has a variety of themed and group tours. Check the website for details.

Countdown: Friday–Monday 10:00 a.m.–4:00 p.m. (last entry)

Mission budget: $$

Flight team: 510-521-8448 | https://uss-hornet.org

Coordinates: 37.7727, -122.3027 | 707 West Horney Avenue, Alameda CA, 94501

Flight plan: Located on the waterfront in western Alameda, the museum has free parking on-site.

Rations: Plenty of restaurants can be found in Alameda and elsewhere throughout the East Bay.

Aerospace Museum of California, Sacramento CA

Debriefing: The Aerospace Museum of California is an interactive facility combining educational STEM exhibits with hundreds of artifacts related to aviation and space travel. The museum has an outdoor aviation park with over thirty-five aircraft. Given the location, at the site of the former McClellan Air Force Base, a fighter jet facility, the museum includes many fighters like the F-4 Phantom, F-80 Shooting Star, A-10 Thunderbolt, and others. One unique artifact is a Makani M600 energy kite, an airborne wind-turbine concept vehicle.

The STARS: One focus of the museum's space offerings are interactive exhibits about the evolution of engines, from piston engines to space-age rocket engines, some of which are on display. Another exhibit imagines Mars in 2034, offering a futuristic look at a hypothetical crewed mission to the Red Planet.

EVAs: Open on Saturdays from 11:00 a.m. to 3:00 p.m., the Flight Zone is a learning laboratory with multiple digital flight simulator stations. Participants can try a variety of simulations under the guidance of an experienced flight instructor. Designed for 5th–12th graders, the program is also open to the general public. Tickets are required and more information can be found on the museum website.

Countdown: Thursday–Sunday 9:00 a.m.–4:00 p.m.

Mission budget: $$

Flight team: 916-643-3192 | https://aerospaceca.org

Coordinates: 38.6751, -121.391 | 3200 Freedom Park Drive, McClellan CA, 95652

Flight plan: Located in McClellan Park, northeast of downtown Sacramento, the museum has free parking on-site.

Rations: The on-site **Old Crow Cafe** offers breakfast and lunch when the museum is open.

Evergreen Aviation and Space Museum, McMinnville OR

Debriefing: The centerpiece of this museum is the original *Spruce Goose*, the largest wooden airplane ever constructed. Other aviation highlights include a number of military helicopters and fighters, an SR-71 Blackbird, and a replica 1903 *Wright Flyer*.

An impressive space exhibition includes several original objects and a wide variety of replicas, many of which are on loan from, or were built by, the Kansas Cosmosphere and Space Center. For this reason, Evergreen has several significant items on display from the Soviet space program.

The STARS: Space highlights include an unflown Project Mercury capsule, a replica Apollo command module used in the splashdown scenes of the film *Apollo 13*, and a replica Gemini spacecraft used in the HBO miniseries *From the Earth to the Moon*.

Other replicas include a lunar module, a lunar roving vehicle, and the 2004 *Mars Spirit* rover. There's a prototype X-38 crew return vehicle and a trainer Skylab Airlock from 1973. A pair of space-relevant planes includes an actual Talon T-38 trainer and a mock-up North American X-15, the experimental rocket plane that eight test pilots flew over 50 miles high to earn their astronaut wings.

Rounding out the displays is a series of items from the Soviet space program. (Because many of these items are on loan from the Cosmosphere, check the Evergreen website or call ahead to see what is currently on display.) Included is a Foton 6 capsule, an unmanned version of the Vostok capsule, which launched the first human into space, Yuri Gagarin, in 1961.

A replica Vega space probe, two of which flew past Venus in the 1980s with both landers reaching the surface, and one transmitting data for fifty-six minutes before being destroyed by pressure and heat. There is also an RD-107 engine from an R-7 rocket and replicas of the 1957 Sputnik, the first human-built object to orbit the earth, and the *Lunokhod 2*, a Moon lander that traveled 23 miles across the lunar surface during 1973.

EVAS: The museum offers a *Spruce Goose* Cockpit Tour on a first-come, first-served basis.

Countdown: Monday–Thursday 10:00 a.m.–4:00 p.m.; Friday–Sunday 10:00 a.m.–5:00 p.m.

Mission budget: $$

Flight team: 503-434-4180 | www.evergreenmuseum.org

Coordinates: 45.2042, -123.1443 | 500 Northeast Captain Michael King Smith Way, McMinnville OR, 97128

Flight plan: The museum is several miles east of McMinnville, just off OR-18, and free parking is available on-site.

Rations: The museum has a café that's open for lunch Friday through Sunday

Orbital neighbors:

Less than two hours south, in Eugene, the **Oregon Air and Space Museum** is a small aviation-focused museum offering a few space items. **Flight team:** www.oasmuseum.com/

Located near Bend OR, the **Sunriver Nature Center and Observatory** combines eight acres of nature trails, bike paths, animals, forests, wetlands, and gardens with nighttime telescope viewing at the Oregon Observatory. There's also a small planetarium on-site. **Flight team:** https://snco.org

---

Museum of Flight, Seattle WA

Debriefing: With over 175 aircraft and spacecraft, the Museum of Flight is one of the largest aerospace museums in the world. Aviation highlights include a British Airways Concorde named *Alpha Golf*, which set a New York-to-Seattle speed record of 3 hours 55 minutes and 2 seconds. Visitors will also find a P-38 Lightning from World War II and reproductions of both the Wright 1902 *Glider* and 1903 *Flyer*. The space exhibition includes objects from the Apollo and space shuttle programs, plus several items from Soviet and Russian space programs.

The STARS: Among the space highlights are an Apollo command module, CM 007A, the first testing and training capsule delivered to NASA for the Moon program. There are mock-ups of the *Apollo 17* lunar ascent stage and a lunar roving vehicle. A display of F-1 rockets, which launched *Apollo 12* and *16*, includes two that were raised, after forty-six years, from the bottom of the Atlantic Ocean.

There's also a full fuselage trainer, a full-scale mock-up of the space shuttle *Orbiter*, without wings, which was used for training. The Viking lander flight capsule is identical to the two that launched in 1975 and landed on Mars eleven months later.

The actual Resurs 500—a Vostok-derived capsule launched by the Russian Federation in 1992 as their first commercial space vehicle—which commemorated the 500th anniversary of Christopher Columbus's voyage to the Americas. Another Russian artifact is the Soyuz TMA-14, a fourth-generation spacecraft that docked on the International Space Station (ISS) in 2009.

EVAs: The museum offers a series of tours and immersive experiences. Check the website for current offerings.

Countdown: Thursday–Monday 10:00 a.m.–5:00 p.m.

Mission budget: $$$

Flight team: 206-764-5700 | www.museumofflight.org

Coordinates: 47.5187, -122.2961 | 9404 East Marginal Way South, Seattle WA, 98108-4097

Flight plan: The museum is in southern Seattle, not far from I-5, with free parking on-site.

Rations: The **Wings Café** is the museum's on-site restaurant.

Orbital neighbors:

Thirty miles north of Seattle, the **Boeing Future of Flight Aviation Center** is an interactive museum focused on the company's efforts in aerospace. Much of the center focuses on Boeing aircraft, but there are some space exhibits, including a full-scale mock-up of the ISS Destiny module. Flight team: www.boeingfutureofflight.com

Located about 100 miles east of Portland OR, **Goldendale Observatory** is a Washington state park with one of the world's largest public telescopes. Afternoon programs focus on the Sun, while nighttime programs observe stars and other space objects. Flight team: www.goldendaleobservatory.com

Over the Canadian border in Vancouver, British Columbia, the **H. R. MacMillan Space Centre** is an astronomy and space museum with plenty of highlights. Exhibits span the search for alien life, exploring Mars, and *Apollo 11*—including astronaut Michael Collins's spacesuit. There's an observatory with a public telescope, a theater with live space science shows, and a planetarium with star shows. Flight team: www.spacecentre.ca.

86. A Hasselblad camera is mounted on the spacesuit of astronaut Alan Bean, who takes a soil sample on *Apollo 12*. NASA.

APPENDIX OF RESOURCES

Books

There are too many books to list! But here are a few that may appeal to readers. If you're looking for a single book that covers the bulk of the space age, I suggest *Spaceflight: The Complete Story, From Sputnik to Curiosity* by Giles Sparrow. It's a beautifully illustrated encyclopedia of space travel, with articles about missions and events, color photos, and detailed diagrams of spacecraft and rockets. The second edition was published in 2019, with a foreword by Buzz Aldrin.

A comparable book is *The Smithsonian History of Space Exploration: From the Ancient World to the Extraterrestrial Future* by Roger D. Launius. Here is another comprehensive volume that includes color photos (but no diagrams), and it goes into a little more written detail about each topic.

A similar book, more focused on the near future, is *Space 2.0: How Private Spaceflight, a Resurgent NASA, and International Partners are Creating a New Space Age* by Rod Pyle with a foreword by Buzz Aldrin.

Magazines and Websites

At time of research, Smithsonian's *Air & Space Magazine* was available as a quarterly print magazine to members of the museum's National Air and Space Society, which carries additional benefits such as event invites and discounts. Meanwhile, the Smithsonian continues to publish many space-related articles on its magazine website (www.smithsonianmag.com/category /space/). This is an excellent source for space news, in-depth features, and ongoing coverage of relevant topics.

Another excellent website is www.space.com, covering space and astronomy news and other content. Meanwhile, NASA regularly posts updates and short articles about agency efforts on its website: www.nasa.gov. Of course, space is such a popular topic that many other news websites and online and/or print magazines cover the topic as well, including www.cnn.com, www.wired.com, and more.

Documentaries and Television Shows

There are so many out there, but one beloved program is *NOVA*, broadcast on PBS. It's broadly focused on science but with many excellent episodes related to space. Similarly, the PBS show *Cosmos* was well known for exploring space topics. It was originally hosted by Carl Sagan in the 1980s and rebooted in 2014 with thirteen excellent episodes hosted by Neil DeGrasse Tyson. A pair of popular feature-length documentaries using mostly archival footage of the historic Moon landing are *For All Mankind* (1989) and *Apollo 11* (2019).

On the docudrama front, *From the Earth to the Moon* (1998) is an HBO miniseries with twelve parts dramatizing the story of the Apollo program. Meanwhile, a unique fictionalization is the alternate history *For All Mankind* (2019) on AppleTV+. It reimagines the post-1969 space age if the Soviets had won the Moon race.

Films

Once again too many to list, but here are a few popular space films that are mostly rooted in fact or attempt to authentically depict contemporary space travel. On the historically accurate side, *The Right Stuff* (1983), based on the Tom Wolfe book, dramatizes the early days of NASA, including breaking the sound barrier and Project Mercury. *Apollo 13* (1995) recounts the in-space rescue of the damaged spacecraft. *Hidden Figures* (2016) tells the story of Katherine Johnson and two other African American NASA mathematicians working behind the scenes on Project Mercury. *First Man* (2018) dramatizes the life and career of Neil Armstrong, leading up to the first Moon landing.

On the science fiction side, here are some films that attempt to depict space travel in an authentic way, imagining near-future scenarios. *2001: A Space Odyssey* (1968) follows a mission to Jupiter to unravel the mystery of an extraterrestrial monolith. *Gravity* (2013) imagines an in-orbit disaster befalling two astronauts separated from a space shuttle. *The Martian* (2015) depicts a NASA astronaut trying to survive after being stranded on Mars. Other films recognized for realistic depictions of space travel include *Interstellar* (2014) and *Ad Astra* (2019).

BIBLIOGRAPHY

Benson, Tom. "Brief History of Rockets." National Aeronautics and Space Administration. Accessed May 31, 2021. https://www.grc.nasa.gov/www /k-12/trc/Rockets/history_of_rockets.html.

Billings, Linda, ed. *50 Years of Solar System Exploration: Historical Perspectives.* Washington DC: National Aeronautics and Space Administration, 2021.

Boyle, Rebecca. "How Was the Moon Made? We Won't Know Until We Go Back." *New York Times*, June 10, 2019.

Chaikin, Andrew. *Space: A History of Space Exploration in Photographs.* Buffalo NY: Firefly, 2004.

Disneyland. "Man in Space," "Man and the Moon," and "Mars and Beyond." Created by Walt Disney Company, 1952, 1953, 1956.

Eisenhower, Dwight D., and Nikita Khrushchev. "Remarks of Welcome to Chairman Khrushchev of the USSR at Andrews Air Force Base." American Presidency Project. Accessed September 15, 1959. https://www.presidency .ucsb.edu/documents/remarks-welcome-chairman-khrushchev-the-ussr -andrews-air-force-base.

Finney, John W. "Washington Is Surprised by President's Proposal." *New York Times*, September 21, 1963.

Gainor, Chris. *To a Distant Day: The Rocket Pioneers.* Lincoln: University of Nebraska Press, 2008.

Gallentine, Jay. *Infinity Beckoned: Adventuring through the Inner Solar System, 1969–1989.* Lincoln: University of Nebraska Press, 2016.

Glover, Ken. "Apollo Lunar Surface Journal." National Aeronautics and Space Administration. Accessed June 5, 2018. https://history.nasa.gov/alsj.

"How Man Will Meet Emergency in Space Travel." *Collier's*, March 14, 1953.

"Image Galleries." National Aeronautics and Space Administration. Accessed January 4, 2022. https://www.nasa.gov/multimedia/imagegallery/index .html.

Keller, Bill. "Eclipsed." *New York Times Magazine*, June 27, 1999.

Kennedy, John F. "Address Before the 18th General Assembly of the United Nations." John F. Kennedy Presidential Library. Accessed September 20, 1963. https://www.jfklibrary.org/archives/other-resources/john-f-kennedy -speeches/united-nations-19630920.

Launius, Roger D. *The Smithsonian History of Space Exploration: From the Ancient World to the Extraterrestrial Future*. Washington DC: Smithsonian, 2018.

Leonov, Alexei. "The Nightmare of Voskhod 2." *Air & Space Magazine*, January 2005.

Logsdon, John, ed. *The Penguin Book of Outer Space Exploration: NASA and the Incredible Story of Human Spaceflight*. New York: Penguin, 2018.

"Man in Space." *New York Times*, April 16, 1961.

Mare Tranquillitatis. USGS Astrogeology Science Center. Accessed October 18, 2010. https://astrogeology.usgs.gov/rpif/videos/making-craters.

Meltzer, Michael. *When Biospheres Collide: A History of NASA's Planetary Protection Programs*. Washington DC: National Aeronautics and Space Administration, 2011.

"More about Man's Survival in Space." *Collier's*, March 7, 1953.

"NASA History." National Aeronautics and Space Administration. Accessed January 2, 2020. https://www.nasa.gov/topics/history/index.html.

"NASA History Division." National Aeronautics and Space Administration. Accessed January 25, 2021. https://history.nasa.gov.

Phinney, William C. *Science Training History of the Apollo Astronauts*. Washington DC: National Aeronautics and Space Administration, 2015.

Ryan, Cornelius. "World's First Space Suit." *Collier's*, February 28, 1953.

Sagan, Carl. *Pale Blue Dot: A Vision of the Human Future in Space*. New York: Random House, 1994.

Siddiqi, Asif A. *Beyond Earth: A Chronicle of Deep Space Exploration, 1958–2016*. Washington DC: National Aeronautics and Space Administration, 2018.

———. *Challenge to Apollo: The Soviet Union and the Space Race, 1945–1974*. Washington DC: National Aeronautics and Space Administration, 2000.

"Skylab's Fiery Fall." *Time*, July 16, 1979.

Sparrow, Giles. *Smithsonian: Spaceflight: The Complete Story, From Sputnik to Curiosity*. New York: DK, 2019.

"V-2 Rocket Attacks and Defense." General Board, United States Forces, European Theater Antiaircraft Artillery Section, 1945. https://usacac.army.mil/sites/default/files/documents/carl/eto/eto-042.pdf.

Von Braun, Wernher, and Willy Ley. "Man on the Moon." *Collier's*, October 18, 1952.

Von Braun, Wernher, Willy Ley, Fred L. Whipple, Joseph Kaplan, Heinz Haber, and Oscar Schachter. "Man Will Conquer Space Soon." *Collier's*, March 22, 1952.

Von Braun, Wernher, and Cornelius Ryan. "The Baby Space Station: First Step in the Conquest of Space." *Collier's*, June 27, 1953.

Von Braun, Wernher, Fred L. Whipple, and Willy Ley. "More about Man on the Moon." *Collier's*, October 25, 1952.

Von Braun, Wernher, Fred L. Whipple, and Cornelius Ryan. "Is There Life on Mars?" *Collier's*, April 30, 1954.

Woods, David. *Apollo Flight Journal*. National Aeronautics and Space Administration. Accessed November 11, 2019. https://history.nasa.gov/afj.

Zak, Anatoly. "Alexei Leonov's First Spacewalk Wasn't Quite as Dramatic as We Thought." *Air & Space Magazine*, March 26, 2020.

CPSIA information can be obtained
at www.ICGtesting.com
Printed in the USA
LVHW041122070423
743512LV00001B/1